Richard Raucci

Mosaic™
for Windows®

A hands-on configuration and set-up guide to popular Web browsers

With 185 Illustrations

Springer Science+Business Media, LLC

Richard Raucci
435 Eureka Street
San Francisco, CA 94114 USA
rraucci@well.com
rraucci@interramp.com
http://www.well.com/Community/rraucci/raucci.html

Mosaic is a registered trademark of the National Center for Supercomputer Applications at the University of Illinois, Urbana–Champaign.
Netscape is a registered trademark of Netscape Communications Corporation.
Windows is a registered trademark of Microsoft Corporation.

Library of Congress Cataloging-in-Publication Data
Raucci, Richard.
      Mosaic for Windows : a hands-on configuration and set-up guide to popular Web browsers / Richard Raucci.
        p. cm.
      Includes bibliographical references.
      ISBN 978-0-387-97996-0   ISBN 978-1-4615-9672-1 (eBook)
      DOI 10.1007/978-1-4615-9672-1
      1. Mosaic (Computer file) 2. World Wide Web (Information retrieval system) 3. Internet (Computer network) I. Title.
TK5105.882.R36 1995
025.04–dc20   95-11636

Printed on acid-free paper.

Production managed by Bill Imbornoni; manufacturing supervised by Jacqui Ashri.
Typeset by Baloo Typesetting, New York, NY.

9 8 7 6 5 4 3 2 1

This one is dedicated to Liz and Philip.

Acknowledgments

I would like to thank Martin Gilchrist, of Springer-Verlag, for his belief in this project. I would also like to thank the Exploratorium museum (http://www.exploratorium.edu) staff and volunteers for giving me the ability to help the public learn Mosaic; I learned a lot myself that way. Lisa Stapleton, now at InfoWorld, was particularly helpful in showing what to expect from a computer book contract. The WELL in Sausalito (http://www.well.com) provided me with countless hours of stimulating exchange with a vibrant online community; the information and tips I got there was invaluable. I would also like to thank the vendors involved, especially SPRY, Spyglass, Netmanage, ISDN*Tek, and Motorola, for giving me speedy access to review software and hardware. David Spolestra of Truevision was invaluable for his perceptive critical assessment of the manuscript.

Contents

Acknowledgments vii

Introduction   1

1 Getting Started   5

2 Mosaic Web Browsers: A Summary 29

3 The Access Provider 57

4 Modems and ISDN Adapters 69

5 Venturing onto the Net 79

6 Gopher: The Worldwide Internet Encyclopedia 95

7 Files on the Net 111

8 On-Line Magazines, Journals, and Books 123

9 Art, Games, Music, and More 143

10 Serious Productivity 161

11 Closing Considerations 173

Introduction

What is Mosaic? Mosaic is the key to the Information Superhighway. It's the best interface to the Internet available. The Internet, in turn, is the real Info Superhighway—a place to exchange information of all types with people all over the world, find files, search extensive databases, play games, and more.

The Windows PC platform fits into this picture nicely. As a client, a standard Windows PC can use Mosaic well. All it takes is an inexpensive modem, an Internet access provider, and a copy of Mosaic for Windows. This book will show you how to set up and compare a variety of versions of Mosaic and Mosaic-style Internet browsers for the Windows platform, including NCSA Mosaic (the first version, available free over the Internet), Netscape Communication's Netscape (a beta version is available free over the Internet, as well as a limited commercial release; this version is also known as "Mozilla"—the Godzilla monster Internet application), SPRY, Inc.'s AIR Mosaic (available commercially as a separate bundled product from several vendors and as a part of the Internet in a Box package), and Spyglass, Inc.'s Enhanced NCSA Mosaic (available commercially from several sources). We'll also look at Enterprise Integration Technologies' WinWeb and Netmanage's WebSurfer.

Mosaic developed at the National Center For Supercomputing

Applications at the University of Illinois, Urbana–Champaign as a part of the World Wide Web project. The WWW was conceived as a way to manage the vast amounts of information on the Internet. Mosaic was designed to be a Web browser, an application that could load information from a variety of sources (server file directories, text, and images, for example) into a common interface. Previously, Internet information of different types had to be accessed from several different applications, resulting in too many steps for the process to be efficient (for example, you had to download a picture file from a remote site, then open it in a paint program to view it; this meant that you couldn't *see* the image before you downloaded it).

The Web browser Mosaic changed all of this. Finally, indexed multimedia information could be displayed in a common interface. Mosaic not only displays picture and text files internally, it also incorporates a hypertext interface. This means that a Mosaic home page (an Internet document) can contain links to other Internet sites and information. This interface is in a standard point-and-click format. Highlighted images and text contain the links. To continue the example given above, Mosaic can now load in a series of small, thumbnail in-line images for browsing pictures at a remote site, so you can see what the picture looks like before downloading. Each small picture can be linked to a larger version; clicking on the picture will start the download automatically, and Mosaic will even launch an external viewer (when configured properly) to display the big picture.

Great stuff. Also great was that at the time that Mosaic was originally designed, the programmers at NCSA decided that the technology had to be available for all users. Versions for the three main platforms (Windows, Macintosh, and Unix) were developed at the same time by different teams at NCSA, led by Marc Andreesen (a principal architect of the original Mosaic concept). The Windows version grew out of this. Originally written by Chris Wilson and Jon Mittlehauser, WinMosaic started out as a 16-bit application that ran under Windows 3.1. It changed during its Alpha release phase to a 32-bit application for Windows For Workgroups 3.1, Windows NT, and Win 3.1 (with a special subsystem installed). When these developers left NCSA, WinMosaic was handed over to a new development team, and continues to be improved and maintained.

Jon Mittlehauser left to join Netscape Communications, a start-up company led by Marc Andreesen of NCSA and the former CEO of Silicon Graphics, Jim Clark. Four other NCSA Mosaic developers also joined the company, and they developed an independent version of Mosaic known as Netscape ("Mozilla"), including a Windows client.

Chris Wilson joined SPRY, Inc., a developer of networking and Internet access software, and created AIR Mosaic, a Win 3.1-based version of Mosaic available commercially that includes many enhancements to the original product.

Meanwhile, back at NCSA, a visualization software developer named Spyglass entered into a partnership with the Center, resulting in Spyglass's handling licensing agreements for the Mosaic code with third-party developers. This resulted in a version known as Enhanced NCSA Mosaic for Windows available from various sources as an add-on product. This code licensing could also result in the development of even more versions of the Mosaic client, with additional Windows-based features.

What does this mean to the user? On the one hand, the free version of Mosaic is just that—free. It's also closer to the heart of the World Wide Web, and less likely to be developed purely for commercial reasons. On the other hand, commercial versions of WinMosaic have more developers working on them, which can lead to advanced features appearing sooner. These versions can also go through more rigorous testing cycles, meaning that they can come a bit closer to being bug-free (Never? Well, hardly ever!).

1
Getting Started

To get started with Mosaic for Windows, you should first check out your PC's configuration. There are various flavors of the PC version of Mosaic available from the NCSA FTP site and from commercial providers. They all require MS-Windows (no DOS version is available yet). The two main variations of Mosaic for Windows are 16-bit and 32-bit. Standard Windows 3.1 and Windows for Workgroups work in 16-bit mode, as does most Windows programs. The 32-bit Mosaic is for use with Windows NT primarily, although you can run it without NT under Windows 3.1 by using a Win32 subsystem available free from Microsoft (at ftp.microsoft.com).

Look for a version 1.2 or above, with OLE support, to remain current with the Win32 standard.

Why 32-bit?

The 32-bit programs are designed to run better under advanced operating systems available from Microsoft, in the form of Windows NT and Windows 95 (Chicago). These will offer better multitasking (the ability to run more than one program at the same time), and improved crash protection (the tendency for programs to shut down unexpectedly). The catch is that these operating systems are still in the process of being brought to the larger PC market. Although you can run 32-bit programs under Windows 3.1 (by installing Microsoft's Win32 subsystem), it's recommended that you run the standard version of Mosaic instead.

The 16-bit standard version of Mosaic for Windows should also be compatible with Windows 95 and future operating systems from Microsoft.

An added benefit is that it will run on an older 286 in Windows' Standard Mode, although this isn't recommended. The 32-bit version of Mosaic is for 386 systems and up.

FIGURE 1.1.
Icons of 32- and 16-bit NCSA Mosaic versions for comparison.

Third-party commercial versions of Mosaic are also becoming available. These have features that aren't in the free versions, including button bars and integrated viewers. It's nice to have a commercial vendor that will provide you with support, but the free versions of Mosaic available over the Internet are in enough of an advanced state that, with a little bit of configuration work, you can use them with confidence.

How Much PC Do I Need?

Some versions of Mosaic available (these are known as 16-bit) will run under Windows 3.1 on a 286 and up (this means Soft-Windows 1.0 on a Power Macintosh can run the PC version of Mosaic). The recommended PC is a 386 running Extended-Mode Windows 3.1 with at least 4-8 megabytes of RAM for the 32-bit versions, which will also require 32-bit extensions for Windows 3.1. Native 32-bit operating systems, like Windows For Workgroups, Window NT, or Windows 95, can use 32-bit Mosaic version without modifications. Since a crucial part of Mosaic is its

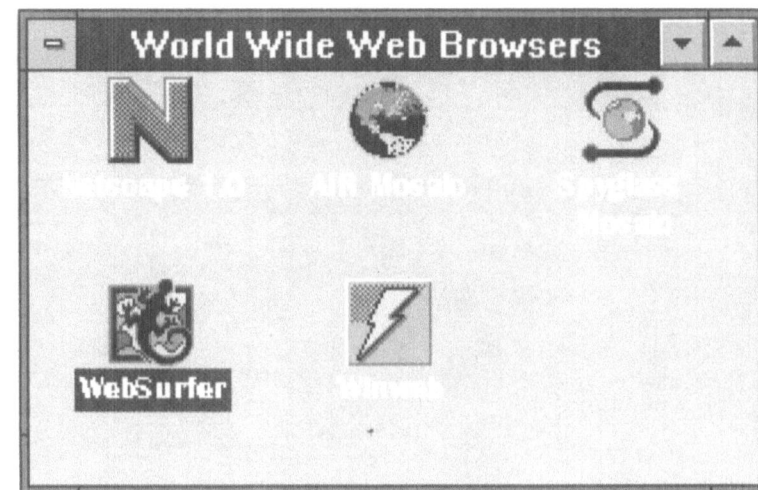

FIGURE 1.2.
SPRY Mosaic,
Netmanage
WebSurfer,
Netscape, WinWeb,
and Spyglass Icons.

multimedia capabilities, a faster system like a 486 (33 MHz) and more RAM (at least 8 megabytes) will do the job better. The extra processing power will help Mosaic and the external viewers it uses to display pictures faster, and sound and animations will play at a more natural rate. Expect your introduction to the multimedia world of the Internet to take a toll on your hard drive as well: sound and picture files downloaded to your PC can take up a lot of space. Note that Mosaic mainly works as a viewer, though; you won't actually be downloading the files that you see under Mosaic unless you want to.

Most importantly, it's crucial that you have Windows set up properly to take maximum advantage of your system resources. An example of a Extended-Mode Windows Memory Control Panel setting for virtual memory is shown in Figure 1.3.

Make sure your system is set up like this, with a proper swapfile (a temporary or permanent section of your hard drive set aside to work with your system's RAM, also know as virtual memory). This will help your system run the various applications that Mosaic uses more efficiently, as well as help Mosaic load picture files into its main screen faster. The way Mosaic displays information from over the Internet means that special consideration needs to be made for the various graphical file formats that Mosaic can handle. Since images are all-important to Mosaic, the quality of color your PC displays under Windows is also very important. Windows' default setting of 16 colors under VGA just won't do it.

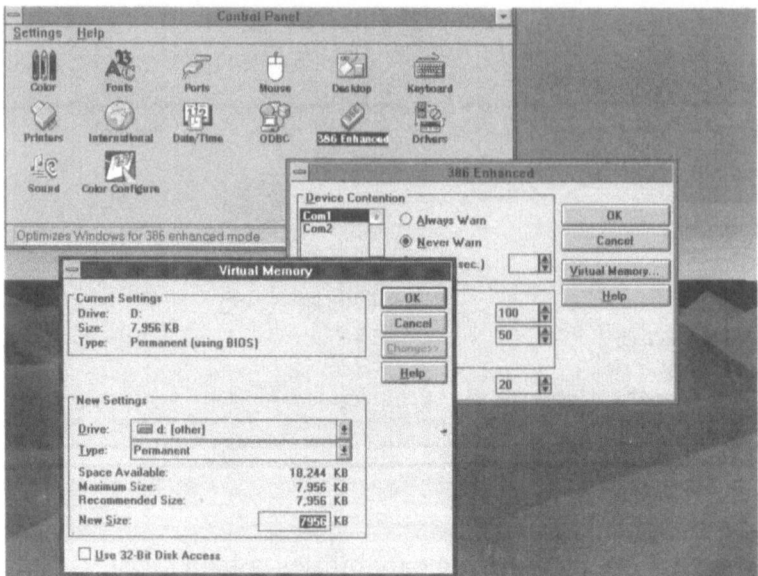

FIGURE 1.3.
Windows 3.1
Memory Control
Panel and swapfile
settings.

This is less than a standard video game machine. Standard pictures on Web pages on the Internet are made up of hundreds to thousands of colors, usually 16- or 24-bit images. The Windows system will convert these images via a process known as dithering to an unusable 16-color palette. You'll be able to see them, but you won't be happy with the results. (See Figure 1.4.)

This isn't the fault of Mosaic. The same picture with Windows properly configured for 8-bit, 256 colors looks like Figure 1.5.

A color setting of 32,000 and above will bring images in even more clearly, but you'll have to make sure your Windows system is up to having a high-color display driver operating. These can slow your system down. Systems with local-bus accelerated video cards could presumably run higher color levels, but the 256-color standard is fine for running Mosaic in most cases.

Note that the display resolution depends on the actual image resolution at the source. It's safe to set Windows at 256 colors as a minimum, and set it for more colors if you're interested in high-end graphics and animations. Increasing the colors in Windows can also affect its overall performance, so setting it for the maximum number of colors your video card will allows is recommended only for faster systems. The 256-color standard is fine for running Mosaic in most cases.

To check your video setup under Windows, click on the Win-

FIGURE 1.4.
16-color 640 × 480
Windows example.

FIGURE 1.5.
800 × 600,
256-color Windows
example.

dows Setup icon, and select the video card option. If you've got a generic VGA setup, it will look like this:

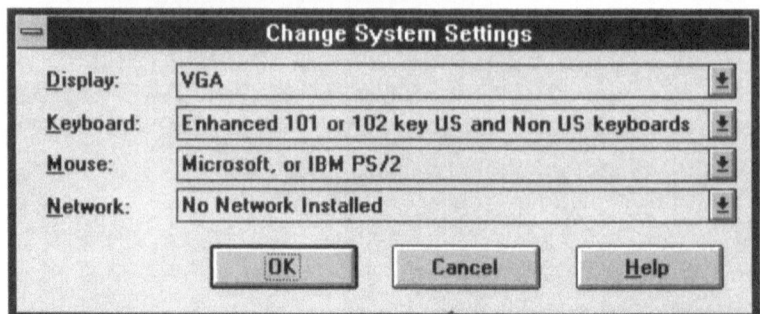

FIGURE 1.6.
Windows Setup with VGA as display driver.

FIGURE 1.7.
Windows Setup with VGA as display driver, Change Systems Settings screen.

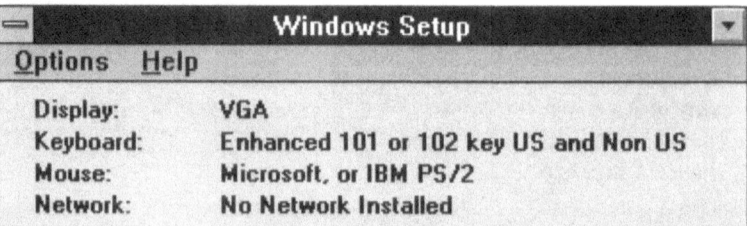

To change the settings, use the Options/Change Systems Settings selection under Windows Setup (in the Main program group). First, make sure your video card is properly installed for Windows. If it is, you'll probably be able to select your video card from a list provided under Windows Setup and Windows will help you install the driver for it. If it's not listed, you'll have to provide the setup disk that came with your video card. Some generic VGA cards may not provide more that 16 colors under Windows; in this case, you'll have to get a newer card in order to run Mosaic properly.

FIGURE 1.8.
Windows Setup with Paradise card installed.

Once you've got the video driver installed, your setup screen will look like this:

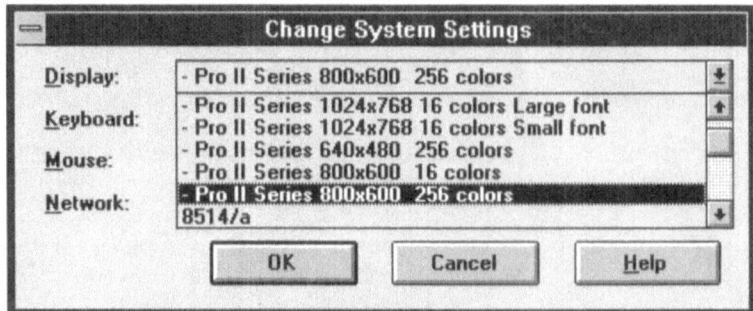

FIGURE 1.9.
Windows Setup with
Paradise example.

The example is for a system with an Orchid Paradise Pro IIs
SVGA card installed. Select the 256 color option. Notice that
next to that option is another set of numbers. This refers to your
monitor's resolution capabilities. A standard VGA monitor is set
at a 640 × 480 resolution:

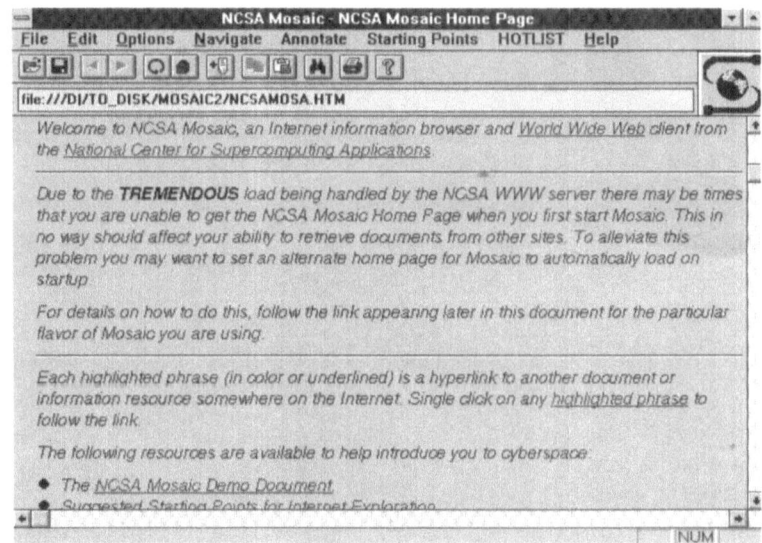

FIGURE 1.10.
NCSA Mosaic at 640
× 480 resolution.

A higher resolution gives you more screen space, and you'll
have the ability to see more windows open at the same time.
Figure 1.11 is a full-screen example of NCSA Mosaic running on
a system set to 800 × 600 resolution.

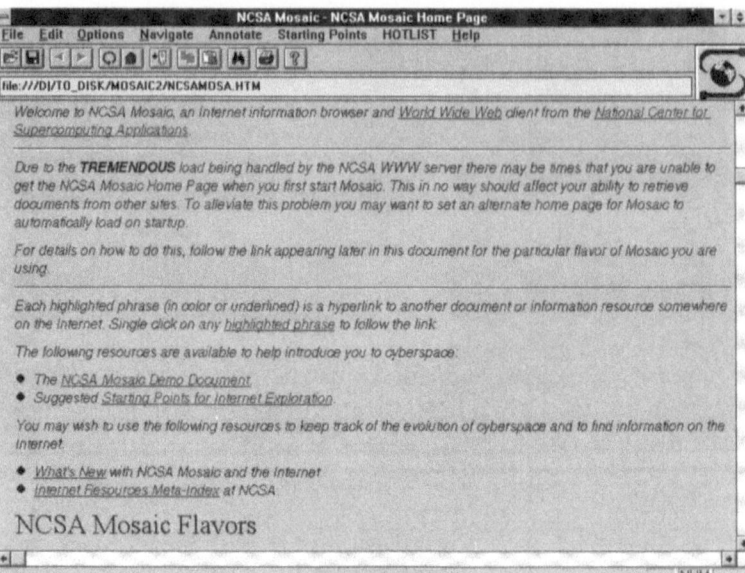

FIGURE 1.11.
NCSA Mosaic at 800
× 600 resolution.

And at 1024 × 768:

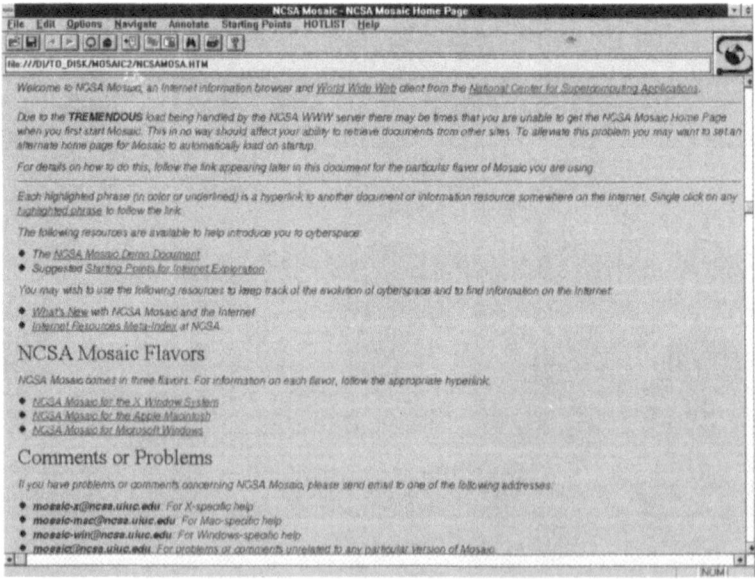

FIGURE 1.12.
NCSA Mosaic
at 1024 × 768
resolution.

Notice how the appearance of Windows also changes. It's not terribly important to run Mosaic at resolutions higher than 640 × 480 in most cases (it's more important to have the colors set properly), so even older VGA monitors should work fine. If you do decide to set your system to a higher resolution under Windows, check your monitor's capabilities first. Make sure your monitor is rated for 800 × 600 in its documentation, for example, or you could run the risk of rendering your display temporarily useless, and you'll have to edit your Windows configuration files to get back to standard 640 × 480. If you want the higher resolutions (the extra screen space is easy to get used to), and your system can't handle it, you'll have to upgrade your monitor.

The different Mosaic for Windows programs don't take up a lot of disk space, from less than 2 megabytes to around five megabytes normally. This is mainly because the program is used as a hypermedia document viewer—the contents come directly off of the Internet into Mosaic and don't take up space unless you decide to save them to your hard drive. You can read a full-color online magazine in Mosaic without having it take up hard drive space, for example. The main exception to this is in Netscape, which caches Web all documents; you'll have to check the /cache directory in your Netscape installation periodically and remove built-up files that are no longer being used.

Where to Get It

There are many ways to get a copy of Mosaic (and other Web browsers). Assuming that you have a modem (preferably at 9600 baud and over), you can start an account with an Internet access provider that will give you the connection you need to access Internet sites like NCSA. You can find Internet access providers listed in computer magazines; it's important to find a current access provider with service in your area (look for one that has a local call connection to the service provider, to avoid long-distance charges).

Make sure your access provider gives you a SLIP (Serial Line Internet Protocol) or PPP (Point To Point) account; these emulate the TCP/IP (Transmission Control Protocol/Internet Protocol) networks that are necessary for transferring information to and from the Internet.

Once you have the Internet access account set up, you will then be able to use a File Transfer Protocol (FTP) program to download the Mosaic software. The FTP program is usually provided in a software package from your access provider. Use it to go to `ftp.NCSA.uiuc.edu` to find the Windows Mosaic program in the `/PC/Windows/Mosaic` directory. You should see a number of files for different versions of Mosaic for Windows, including programs for Windows 3.1. Look for the latest version to have the highest number; for example, the file `mos20a8.exe` in the example illustrated below is a self-extracting archive file for NCSA Mosaic for Windows Alpha 8. You should also read the readme files at the archive site for more information.

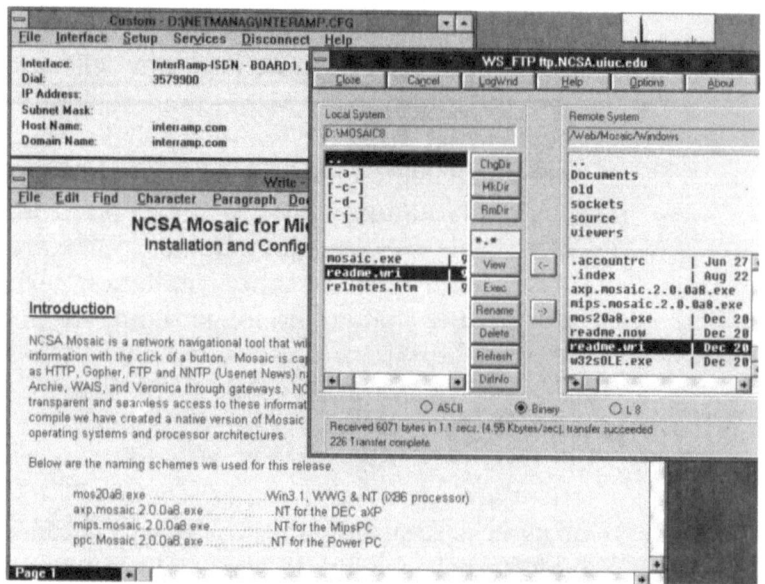

FIGURE 1.13.
Using FTP to access NCSA Mosaic.

If you already have a TCP/IP network connection, you can find out from your network administrator if you'll be able to run FTP, and then you can download the Mosaic program. A good rule of thumb is that if you're able to run FTP, you should be able to run Mosaic.

You can get other Web browsers like Netscape through the same FTP methods as for Mosaic. Netscape for Windows is located at the Netscape Communications servers at `ftp1.mcom.com` and `ftp2.mcom.com` in the `/netscape/windows` directory (also try alternate FTP sites at `ftp.digital.com` and `ftp.barrnet.net` in the `/pub/net/infosys/Netscape` and `/netscape` directories, respectively). EINet's WinWeb browser is located at `ftp.einet.net` in the `/einet/pc/winweb` directory. A demo version of AIR Mosaic Express is available from SPRY, Inc. at `ftp.spry.com` in the `/AirMosaicDemo` directory.

Some commercial Web browsers are available as a part of complete Internet software packages. AIR Mosaic is also available as a part of SPRY's Internet in a Box, which includes Internet access software, connection tools, and trial Internet accounts. Netscape's WebSurfer is a part of the Internet Chameleon package, which also includes Internet connection tools and trail accounts. These packages are good for users with modems, since they provide the connection software (with SLIP or PPP trial Internet access accounts) and bundled Internet programs like FTP and Web browsers.

Remember, you're not limited to the Web browsers that come with one of these packages; you can always use FTP to get a copy of NCSA Mosaic or Netscape instead, and it should work fine as a replacement (as long as you configure it properly). You can also use a commercial Web browser with your Internet connection, once you have established it first.

These commercial versions of Mosaic are available from software companies directly or as a part of book packages (for example, Enhanced NCSA Mosaic from Spyglass is a part of the Mosaic Handbook available from O'Reilly and Associates). It's important to note that without an Internet access account (with a SLIP or PPP connection) you won't be able to use these on the World Wide Web. Make sure you have all of the pieces you need to make your Mosaic Web browser work.

Viewers

Once you've got a version of Mosaic on your Windows system, the first thing to do is to configure file viewers and sound/movie players for it. Mosaic can display small to mid-sized pictures in its main viewing area (these are known as "in-line graphics"), but these small pictures can be linked to full-size images as well. Clicking on one of these will bring the picture information across the Internet to your system, whereupon Mosaic will launch an external viewer so you can see the whole picture. You'll have to configure the viewer for your system, however.

There are a number of ways to do this. The NCSA version of

Mosaic has a control file called `mosaic.ini`. It contains information that Mosaic uses to run, including the information it uses
to launch external viewers and the types of files they can handle.
Editing this file is about as simple as editing Windows 3.1 `win.ini`
and `system.ini` files; if you have some familiarity with them, you
shouldn't have any problems. Most commercial Mosaic versions
include menus and control panels for configuring their viewers.
These make it very easy to set up your Mosaic system.

NCSA provides a configuration/viewer setup panel in 32-bit
Mosaic Alpha 9 and above, or you can use a shareware utility
developed by Rod Potter at York University for earlier versions.
Called SetMosaic (The Unofficial Mosaic Configuration Sheet),
it provides a graphical interface to NCSA Mosaic's `mosaic.ini`
control file. You can find it at the `ftp://ftp.wustl.edu` archive.

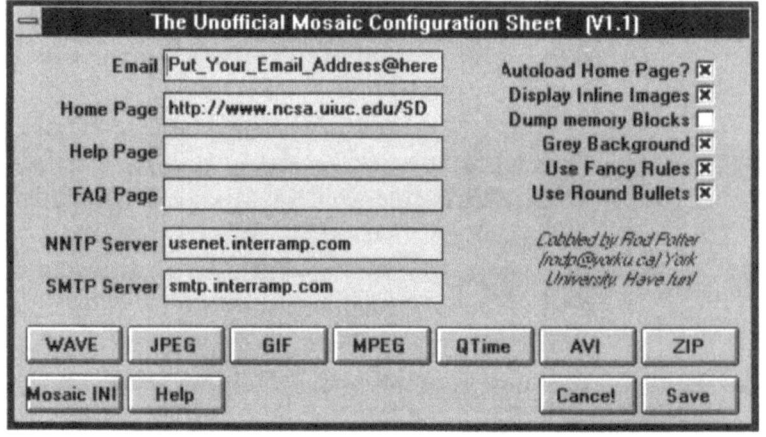

FIGURE 1.14.
SetMosaic, The
Unofficial Mosaic
Configuration Sheet.

A popular shareware external viewer for the Windows platform is a program called LView. LView can display GIF, JPG,
TGA (Targa), and BMP (Windows Bitmap) file formats, and can
be added to your Mosaic program by installing it to your system and adding a pointer to it in your `mosaic.ini` control file.
Once you've brought the picture to your system and displayed it
in the viewer, you can save it to disk. A tip for Windows users:
use LView's Save As command (under the File menu) to convert
loaded GIFs and JPG from across the Internet to BMP files; place
these in your Windows directory, and you can use them as Windows wallpaper backdrops.

LView is also known as ImageViewer for Internet in a Box,

FIGURE 1.15.
An example of
Mosaic at the
Purdue Weather
Server home page,
launching an
external viewer to
load a larger view
of a comprehensive
weather map.

and it's included. For other versions of Mosaic that don't include it, you can find a copy of it at many PC software FTP archives. The Windows Mosaic home page at NCSA provides a link to the program in the Viewers section.

Next to picture information, proper sound configuration under Mosaic is also important. Mosaic for Windows comes pre-configured for Windows' own .WAV audio file format (clicking on one of these will launch the Sound Recorder application for playback), but that's not what's mostly out there on the Internet. NCSA recommends Wplany (Windows Play Any) and WHAM (Windows Hold And Modify), shareware sound players that support standard sound cards and a wide range of sound file formats. WHAM also provides extensive editing capabilities and shows a nice graphic display of the loaded sound file.

You add sound players the same way as you add picture viewers. Note that the standard way of referring to Mosaic's external helper programs is to call them "viewers," regardless of what they actually do; most versions of Mosaic will have you add sound players using their "add viewers" commands. A sound card at the 8-bit, SoundBlaster level is strongly recommended, but you can also download a speaker driver from Microsoft that will allow you to install your PC's internal speaker for use under Windows. Don't expect much from this, however (a PC's internal speaker

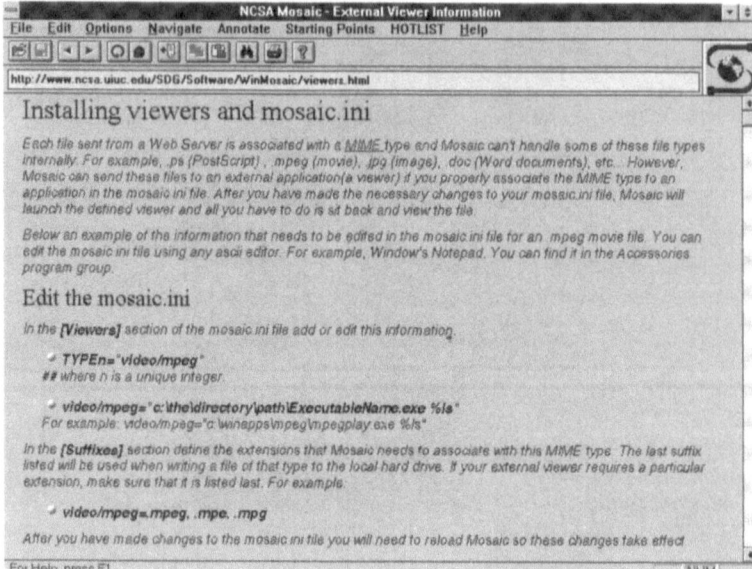

FIGURE 1.16.
NCSA's Viewers
page, with links to
common viewers.

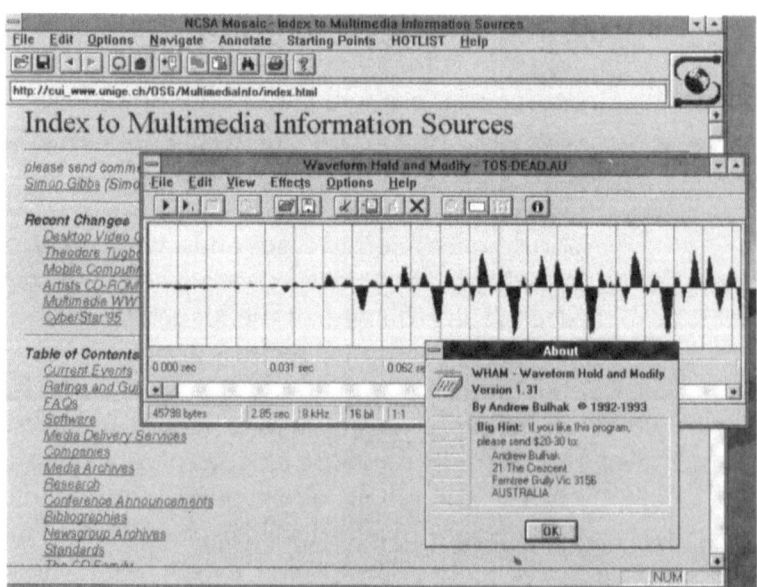

FIGURE 1.17.
WHAM running a
sound file.

is a 1-bit buzzer, not a real speaker), and it can cause problems running larger sound files, including unacceptable screechy sound and locking up Windows. The Microsoft speaker driver is at `ftp.microsoft.com`, and you can get there from the NCSA WinMosaic Viewers pages as well.

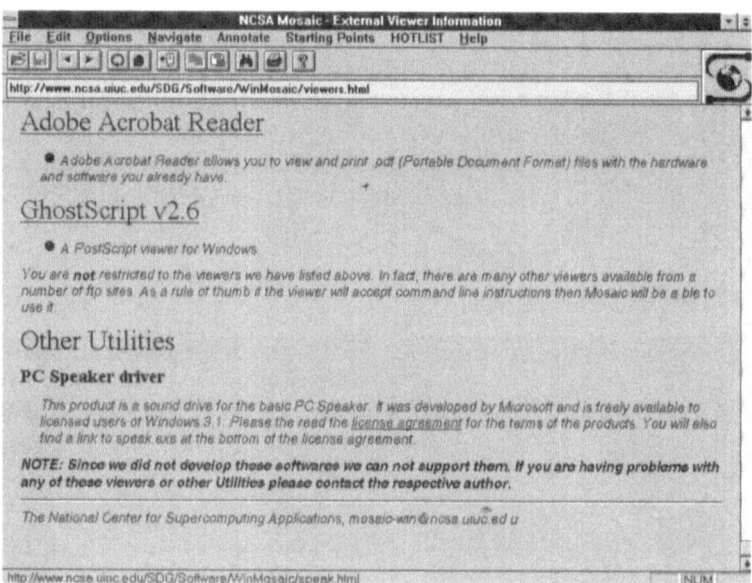

FIGURE 1.18.
Microsoft's PC speaker driver link at NCSA.

You can also play animations and video from over the Internet in standard formats with external viewers for Mosaic. MPEG, the Motion Picture Experts Group standard for compressed video, is widely used on Internet sites that feature film clips and animation. The shareware MPEG viewer MPEGPLAY, developed by Michael Simmons, requires Windows NT or the Win32 subsystem. This viewer is listed as one you can use with Mosaic at the NCSA site, and a link to it is provided on the Viewers page. The unregistered version won't play any file larger than 1 megabyte, however. It's also interesting to note that the compressed ZIP file that MPEGPLAY installs from contains the Win32s subsystem that you'll need to run 32-bit programs under Windows 3.1. You can add this MPEG player and upgrade your system at the same time (but check the Win32 version carefully; NCSA Mosaic Alpha 8, for example, needs Win32 1.2 with OLE, not Win32 1.1).

An alternate MPEG player that works well under Windows (and has no file size limitations in the unregistered version) is

VMPEG, by Stefan Eckart. This also requires the Win32s subsystem. VMPEG is available from `ftp.netcom.com` in the `/pub/cfogg/mpeg` directory, and at selected PC shareware archives. VMPEG also works with the WinG subsystem designed by Microsoft for high-speed animation under Windows (although the ZIP file also installs a non-WinG version). You can find WinG at `ftp.wustl.edu` in the `/pub/MSDOS_UPLOADS/WinG` directory.

Be aware that video on a PC requires a lot of horsepower, and that file sizes for animations and film clips can be very large.

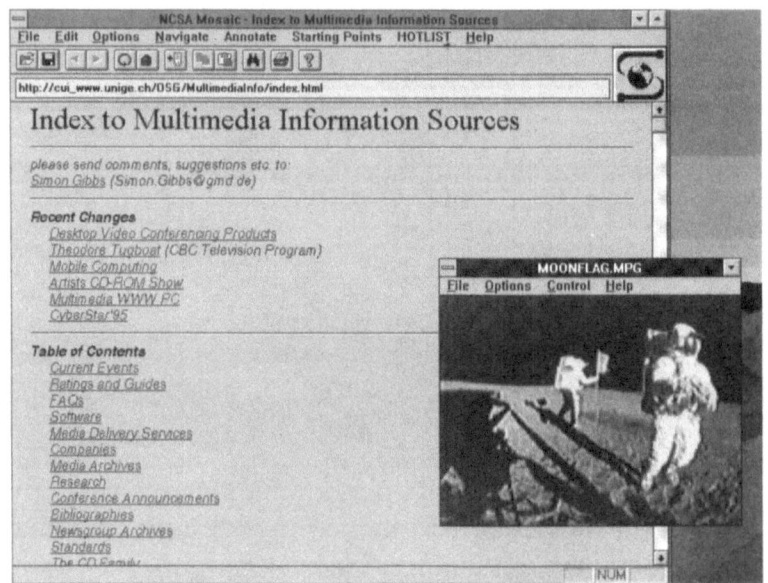

FIGURE 1.19.
VMPEG running an MPEG video clip.

Other video formats include Microsoft's Video For Windows and Apple Computer's QuickTime for Windows. These can also be installed and added to your Mosaic configuration file. QuickTime for Windows is available from Apple's FTP sites (`ftp.apple.com`), and Video For Windows is available from Microsoft (`ftp.microsoft.com`). NCSA also makes available on their Viewers page a link to a subsystem that allows QuickTime movies to be played under the Windows Media Player (the same system that runs Video For Windows AVI files). QuickTime for Windows can also play AVI files. This means that you shouldn't need more that one of these to run *both* QuickTime movies and Video For Windows files. It's best to find out which one displays video clips best on your system, and go with that. Note that QuickTime movies have to be in a "flattened" file format for the Windows Player to be able to use them, as opposed to a pure Macintosh QuickTime format (data plus resource fork). The QuickTime for Windows player can generally handle QuickTime files with QT and MOV extensions. QuickTime for Windows 2.0 will not be available on Internet sites, and has to be purchased from Apple or an authorized reseller.

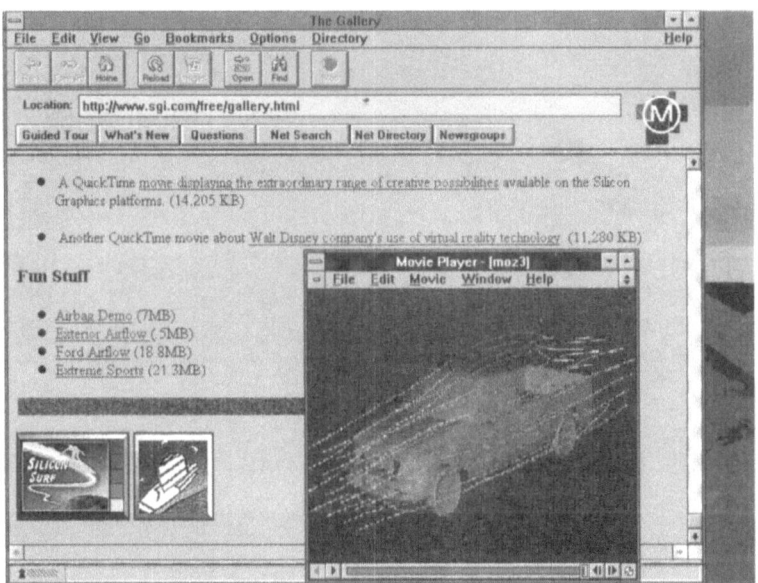

FIGURE 1.20.
QuickTime for Windows 1.1.1 Player running an airflow demo movie.

Video For Windows can be added as an external viewer by specifying it in the mosaic.ini file, or using your Mosaic version's "add viewer" panel. There's no need to load another application to your hard drive; with Video For Windows properly installed, the Windows Media Player will run the AVI files directly.

The PostScript format is popular with many Unix systems that make up a lot of Internet sites, so you may want to look into adding PostScript and Acrobat viewing support to your Windows Mosaic setup. These enable you to see fully formatted documents in a standard format. A popular shareware PostScript extension program for Windows (and Mosaic) is GhostView, which uses a PostScript interpreter called GhostScript. GhostScript consists of a viewer that Mosaic can launch to let you see PostScript files from over the Internet.

GhostScript is available at this point in more of a kit form

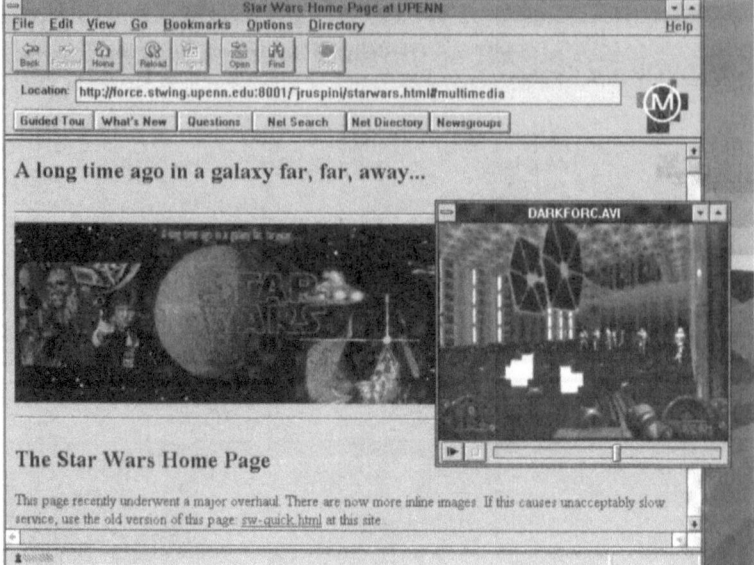

FIGURE 1.21.
Media Player
running a Video For
Windows Star Wars
Dark Forces game
animation.

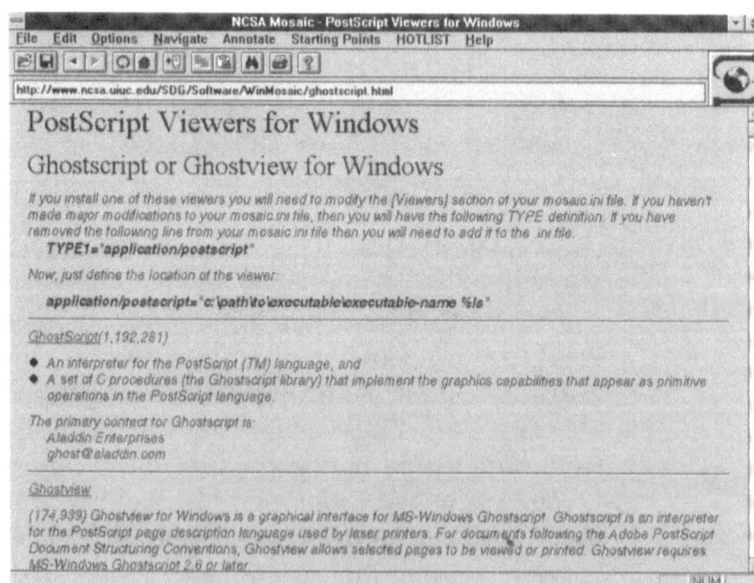

FIGURE 1.22.
NCSA's PostScript
for Windows
page, featuring
GhostScript and
GhostView.

than a finished product. Though it can handle PostScript images well, you'll have to install fonts yourself, or documents with fonts in them probably won't load. The GhostScript interpreter and GhostView for Windows are available through the NCSA Viewers page.

Pure PostScript can be viewed with Adobe Illustrator for Windows, but that's an expensive add-on just to use for a viewer. Adobe developed the Acrobat Portable Document Format as a way to generate files in formats that can be viewed with ease on many different platforms. The Acrobat Distiller converts Post-Script files to the Acrobat format, and Acrobat Reader can be used to read them. You can use Distiller to convert PostScript files you download using Mosaic, but it's also a fairly expensive option for use as a type of viewer.

NCSA provides an introductory page that explains how to download and install the Acrobat Reader for Windows for use with Mosaic. Acrobat Reader is freely distributable. Adobe hopes to make the PDF format more common than the PostScript format for distributed document. Spyglass is working to integrate Acrobat documents directly into Mosaic in a future release. A sample PDF file can be found at the PSI World Wide Web site (`http://www.interramp.com`), an Acrobat version of the PSI User's Guide for their Internet access service. Configuring Mosaic to launch Acrobat Reader to view PDF files is easy, and works well.

Disk space for a set of viewers like the ones outlined above could average out to about 15–25 megabytes or more. It's also important to note that you might want to add these viewers to their own Windows Program Group. They can be used to view picture files and animations you download even without Mosaic running.

Adding viewers may also require that you understand a bit about file formats and disk transferring under Mosaic. You can download these viewers from the NCSA Windows home page site by selecting the Load To Disk option in the Mosaic File Menu, or use Netscape's built-in downloading facility. This will transfer the information from over the Internet to your hard drive, instead of loading it into Mosaic's main screen. Then you can unzip it (most of these viewers for Windows are compressed using PKUNZIP, a DOS/Windows standard file archiver), follow the program's in-

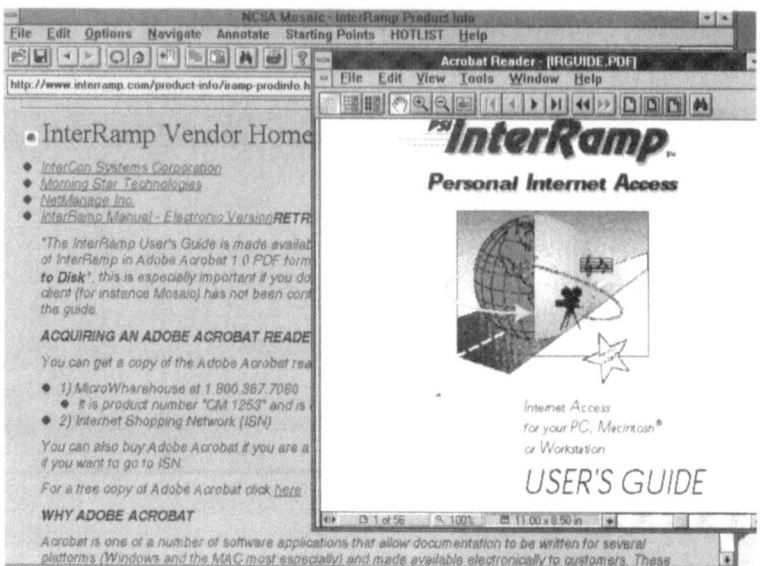

FIGURE 1.23.
Adobe Acrobat
external viewer
displaying PSI's
online User's Guide.

stallation procedure, and add it to the `Mosaic.ini` configuration file. Follow the instructions and load LView first, and you're on your way to using the power of Mosaic to help you equip it to its maximum multimedia potential.

Some of the file formats you'll encounter when using Mosaic are the following:

Image File Formats

.GIF Graphics Interchange Format, a standard PC picture file format, both for pictures viewed inside Mosaic's main screen and as larger, externally viewed files.

.JPG Joint Picture Experts Group compressed image file format, with typically smaller file sizes than with .GIF files. These files can be viewed as in-line graphics in Netscape, or with an external viewer in other versions of Mosaic.

.BMP Windows' Bitmap Paint standard file format, supported by Windows 3.1 Paintbrush and LView.

.PCX Z Graphics standard PC picture file format, also supported by Windows 3.1 Paintbrush.

.TGA Targa (Truevision Advanced Raster Graphics Adapter) file format, supported by LView.

Note: There are many other graphic image file formats in use over the Internet that pertain to specific computer platforms, and that will require you to install a separate program to act as a viewer for files with those extensions. For example, editing your `Mosaic.ini` file to include Photoshop For Windows as a viewer (if it's available on your system) will add a larger variety of image file formats that you will be able to use (like Encapsulated PostScript and TIFF).

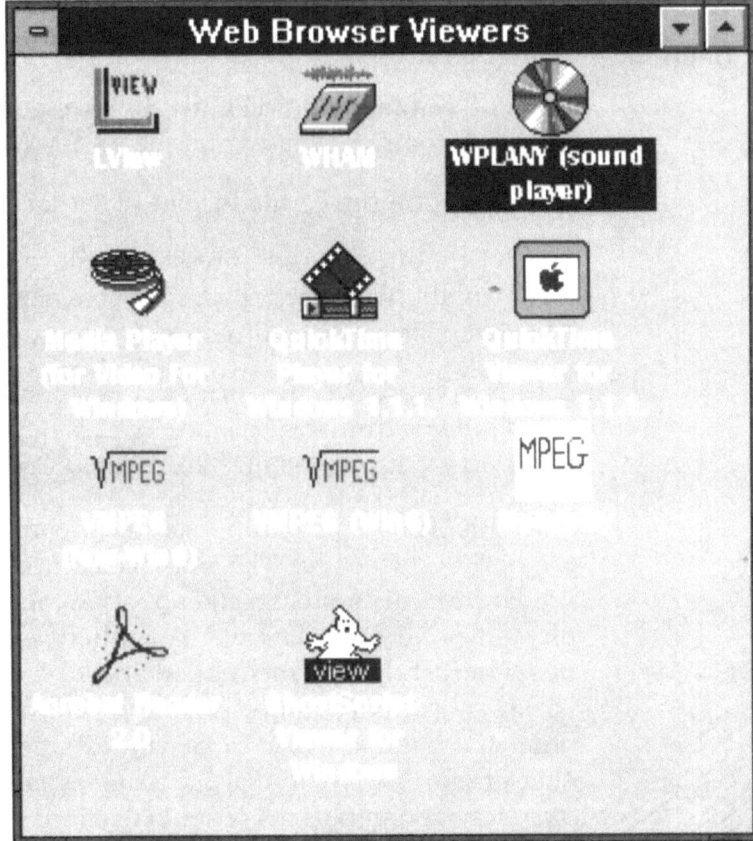

FIGURE 1.24.
Web Browser Viewer's Program Group under Windows.

Motion Video File Formats

.MPG Motion Picture Experts Group compressed motion video file format.

.QT, .MOV QuickTime motion video file format.

.AVI Video For Windows motion video file format.

Note: Motion video files can be very large, and can take a long time to download via Mosaic, especially using a modem. Only certain QuickTime file formats (single-fork "flattened" versions) can be played via QuickTime For Windows.

Compression File Formats

.ZIP Phil Katz's standard PC file compression program, available from many FTP sites and the NCSA Windows home page.

.tar Unix compressed file format (Unix tar command).

.gz GNUtar, alternate compressed tar file format (gzip command), available from the `ftp.univ-rennes1.fr` /pub/Images/ASTRO /anim/soft/msdos directory (gzip program).

.Z Unix compressed file format (Unix uncompress command), available from the `ftp.univ-rennes1.fr` /pub/Images/ASTRO /anim/soft/msdos directory (uncompress program).

Note: PKZIP, the program used to uncompress .ZIP files, is a standard PC file utility. The Unix file compression programs are standard to Unix systems, and are often found as a part of utility software programs (shareware or otherwise) on many different sites (not just the one mentioned above).

The syntax for a compressed file for a Unix Internet site can be somewhat confusing for DOS/Windows users. File systems for other platforms have longer file name capabilities than DOS's 8-letter description plus a 3-letter extension, so it's not uncommon to find files with very long descriptive names. It's usually safe to assume that the last extension is the actual format that the file is in, and the extension immediately *after* the main file name is the format that the unarchived file will be in. Thus, the file: `SolarEclipse.avi.tar.gz` is a Video For Windows AVI file that has been compressed using tar and then gzip. You'll have to run gzip and then tar to extract the file before you can run it. Mosaic can't handle archived files directly, so you'll have to download files like these. Also, remember that DOS will truncate filenames into the (8.3) format, so the file for our example will undergo a name change during the transfer to your system to `Solarecl.gz` or even `Solarecl.avi`, a much less descriptive name (and in the latter case, an inaccurate name. The file is still a .GZ file, and is not now a .AVI file). You should rename it to `Solaravi.gz`, for example, to keep the file contents in line with its name, and to be able to keep track of it if you put it aside to unarchive at a later time.

Encodation File Formats

.UUE UUencoded (Unix to Unix) file format.

.HQX Binary encodation for Macintosh file formats.

Note: Both uuencodation and HQX file formatting make binary files into text files.

UUencoding is used to turn non-text program files into text files that can be transmitted via email or posted as NetNews. Mosaic isn't really designed to use uuencodation (it handles transferring binary files fine without any intermediate steps), but there's a chance you might run into one of these files on an FTP site somewhere down the line. UUencode and uudecode (used to translate uuencoded text files back into binary programs) are utilities that come standard with Unix systems, and are available in DOS shareware versions from FTP sites like `ftp.wustl.edu` (a site with many DOS/Windows shareware files). SPRY's Internet in a Box provides a uuencoding/ decoding utility that runs under Windows, as well.

.HQX formatting is used to convert Macintosh binary files into text files. Mac files typically contain a data fork and a resource fork bundled as a single file; these can get separated during file transfer, resulting in an unusable program. .HQX formatting eliminates this problem, and makes these Macintosh files accessible to any system—including PCs. Typically, the only Mac .HQX files you'll want to download via Mosaic are document files, picture files you're sure your Windows graphics programs can handle, and QuickTime movies. You can convert these .HQX files under DOS/Windows by using a good uudecoding program. Remember that QuickTime movies and picture files have to be in a format that PCs can use; even if they're uuencoded into a single file, they can still decode out to a standard Mac data/resource fork pair that your Windows program can't handle.

Document File Formats

.TXT Standard ASCII text file format (usually, readme files are in ASCII format).

.DOC Microsoft Word document format.

.RTF Rich Text document format, typically with more formatting than ASCII text.

.PS PostScript file format.

.PDF Adobe Acrobat document format.

Note: As with graphics file formats, there are many more document file formats that are specific to certain word processing programs. It is possible, however, to use a single program with a large capacity for translating different file formats as a document file viewer in Mosaic. For example, Microsoft Word 6.0 under Windows can handle almost any word processing file format, except PostScript, including .TXT and .RTF files.

2
Mosaic Web Browsers: A Summary

Mosaic provides a standard interface to the Internet. What it actually looks like is like a standard Windows program; it's as conventional as a Windows word processor.

We'll first look at the NCSA variations. These were developed by the National Center for Supercomputer Applications as a public-domain interface. As such, they're freely available from Internet servers at NCSA. Currently in an Alpha development state, NCSA Mosaic comes in two main versions: Alpha 2, a 16-bit application that runs under Windows 3.1, and Alpha 4 and later versions, 32-bit applications for Windows NT, Windows for Workgroups, and Windows 3.1 (with the Win32 subsystem installed). The Alpha 2 version works without a Win32 subsystem for Windows 3.1, and it's available by using FTP at `ftp.ncsa.uiuc.edu` in the `/Mosaic/Windows/old/wmos` directory.

Alpha 2 was developed by Jon Mittlehauser and Chris Wilson at NCSA in 1993. The main window looks like Figure 2.1.

The large spinning globe icon to the top right of the main window is a progress indicator. It graphically shows how the network transfer of the Mosaic document is progressing as it speeds up and down. When it stops, the document is fully loaded. The globe icon is also a panic button: you can click on it to stop a transfer

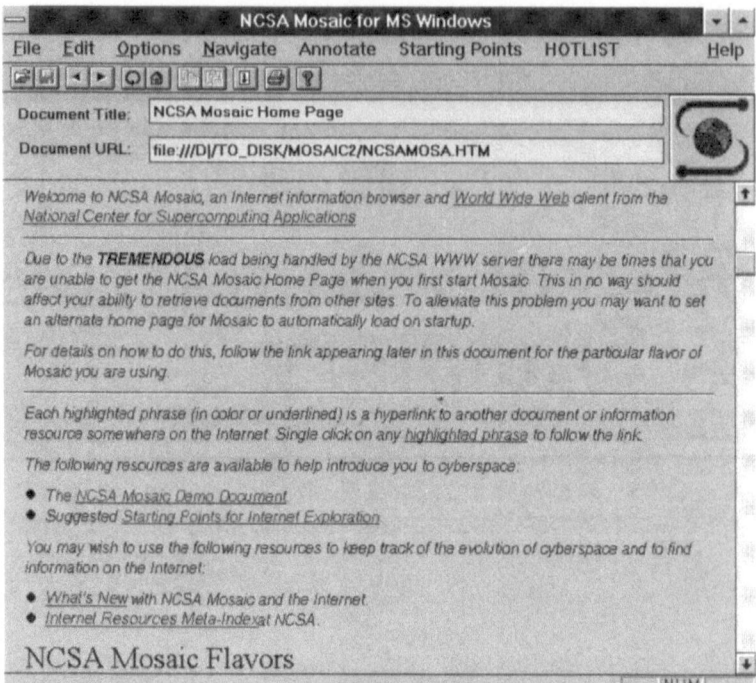

FIGURE 2.1.
The main window for
NCSA Mosaic 2.0
Alpha 2.

that seems to be taking too long, and it will terminate the transfer request.

The Menu bar is familiar to Windows users as a standard program feature. Below that, the program features a set of buttons that correspond to menu items for quick access. The buttons have functions for loading documents and URLs, moving between cached documents (using the front-back arrow keys), and finding text items.

Corresponding to the button bar, the WinMosaic Alpha Two File menu has items for opening Internet URLs and local files. Options to save native HTML documents and to view the document's source file, as well as any printing features, were not implemented for the final public NCSA 16-bit release of Mosaic.

The Edit menu has a Copy item, but no Paste feature, and the only item you can copy is either the Document Title or the Document URL field (you can't copy an item or selection from the document itself). The Edit menu does feature a basic Find dialog box that can search for items in the body of the currently loaded Mosaic document.

The Options menu features a Load To Disk setting, for downloading binary program files that Mosaic can't load in directly. This is a handy feature for using FTP with Mosaic. Other options allow you to show or hide the button bar, Document/URL fields, and the bottom status line. This gives you more space to view a document, but there's a trade-off: the globe icon will disappear if you hide the Document/URL field, and you won't have any indicator of your current document transfer progress. You can also use the Options menu to turn off in-line image display, which will speed up HTML document transfers (but won't show in-line images contained in them). This is best for slow modem connections.

The Navigate menu has the same items for moving forward/backward, document reloading, and accessing a home page as the button bar. It also includes a useful History item, which launches a separate view of where you've been as a list of documents. You can then navigate around your cached documents by clicking on them in the History list, instead of having to move through them one by one forward or backward. This menu also includes an item for adding a document to the Hotlist (discussed below).

Annotations in this version don't seem to be working. These would allow you to add comments to a retrieved document. They're available in the 32-bit Alpha 7 and up releases, however.

The Starting Points menu is one of the key elements to NCSA Mosaic. It includes nested menus for finding out about the World Wide Web, accessing sample home pages and FTP directories, and starting gopher searches. It's really easy to launch into the Internet using these default menus.

The Hotlist is the last item on the menu bar. It consists of a list of documents (Mosaic pages, FTP or gopher sites, and more) arranged by descriptive name. This is a handy feature. After following a link down through several pages and Internet locations, you can save that destination to the Hotlist, where you can access it at any time by scrolling to it on the pull-down list. There's a problem with the Hotlist in WinMosaic Alpha 2: it can get too long, and scroll off the screen. For example, a 1280 × 768 Windows desktop running the program can have a Hotlist that can show only about 35 items, at a default font. You can make the

Hotlist longer than that, but it will display only the first 35, no matter what.

There's no interactive Hotlist editor in this version; the entries are listed in the mosaic.ini file under the <HotList> section. You can edit this using a standard text editor like Notepad. You'll have to use this process to change the default descriptions in the Hotlist as well; otherwise Mosaic gives them the same names as they came in with, and these can be confusing. For example, the Hotlist entry for a directory of electronic magazines at the University of Michigan will look like "FTP directory of /etext.archive.umich.edu/pub/Zines"; you can change this to read "Electronic Magazine Archive at UMich", and it will point to the same site.

The last 16-bit version of NCSA Mosaic, Alpha 2, is a basic Web browser, and does that job well. Unfortunately, it's been left behind in the development cycle, and is missing some basic features of other versions of Mosaic. Its main virtue is that it doesn't need a 32-bit subsystem to be loaded on top of Windows in order to run. On the other hand, Netscape Communication's Web browser is also available in a 16-bit version for standard Windows 3.1, and it has a large amount of features not found in the NCSA Alpha 2 release.

NCSA Mosaic Win32 Versions

The 32-bit versions of NCSA Mosaic for Windows are more fully developed than the Alpha 2 version, but they require a 32-bit operating system like Windows NT, Windows For Workgroups, or Windows 95 (Chicago). You can run this version under Windows 3.1 by installing a compatibility subsystem called Win32, available at no charge from Microsoft (ftp.microsoft.com) and at several sites, including the same FTP server where you'll find Win-Mosaic Alpha 8, ftp.ncsa.uiuc.edu, in the /PC/Windows/Mosaic directory.

Win32 versions present basically the same screen as Alpha 2, but the button bar icons have been colored, and some of them have been changed. The Find icon in Alpha 2 resembles a thermometer over a document, and this has been changed to a set of binoculars. A burning document icon combined with a plus sign

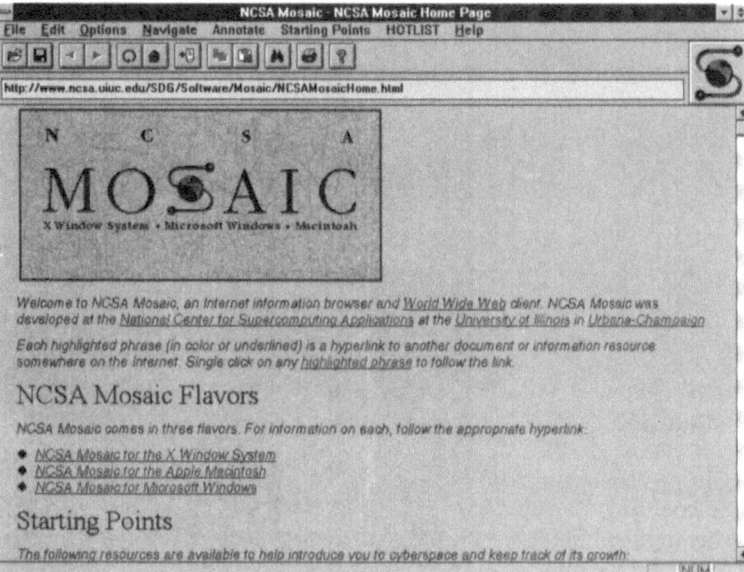

FIGURE 2.2.
The main window for
NCSA Mosaic 2.0
Alpha 8.

is a button used to add a document to the Hotlist, a nice feature. Placing the cursor over one of the buttons pops up a short label indicating its use, a nice touch.

Also changed in these versions was the addition of the document name field to the top window bar, as on a Windows word processor. The swirling globe icon is now linked to the button bar and the URL display line in such a manner that when you choose to hide either of these, the globe becomes smaller, giving you more space to view a document. Unfortunately, hiding both of these still makes the globe disappear entirely, leaving you without a progress indicator.

In the File menu, the same Open, Save, and Print functions now all work correctly. There's also an item for viewing a document's HTML source. This is handy for seeing the actual code construction for the current Mosaic page you're viewing.

Under the Edit menu, the copy and paste functions work, but you still can't highlight and copy any text in the main window.

The Options menu now has an extensive Preferences panel, which you can use to set Mosaic parameters directly, instead of having to edit the mosaic.ini control file. The General section features startup configuration settings you can customize and has a place to enter a default home page. The Services section is where you can add News server information, set the document

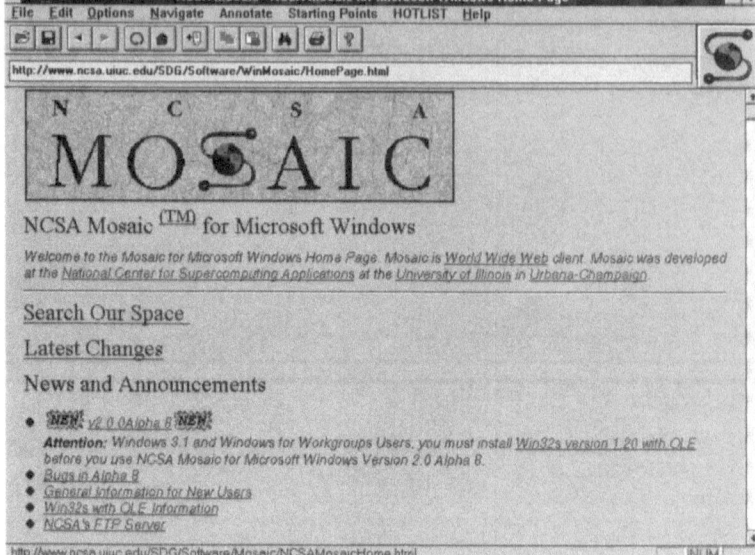

FIGURE 2.3.
NCSA Mosaic 2.0 Alpha 8 Document Source view, showing the HTML document for the Windows Mosaic home page.

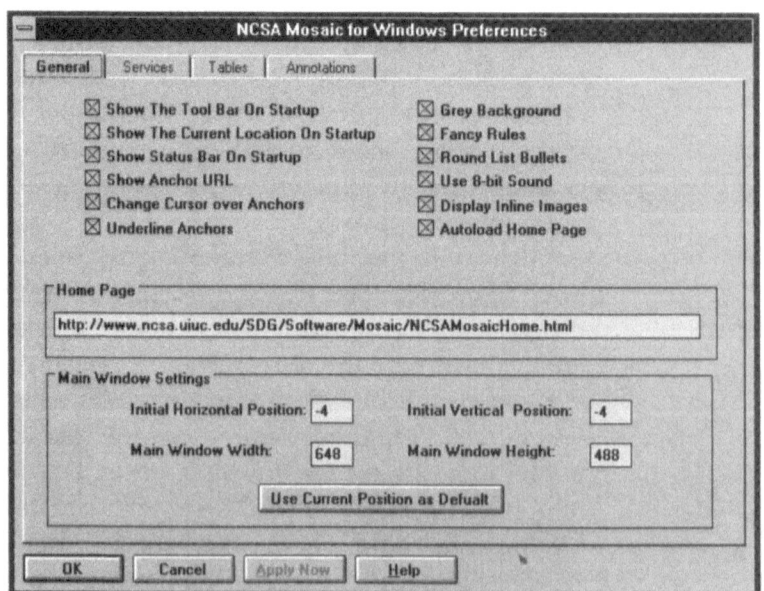

FIGURE 2.4.
NCSA Mosaic Alpha 8's Preferences panel.

cache limit (default is 5 documents), and set up your email address.

Under NCSA Mosaic Alpha 7 and above, the Navigate menu is basically the same as the Alpha 2 release, but the program now features a menu editor. This is one of the best features of NCSA's public version of Mosaic, as it allows you to control your Hotlists. You can now relabel Hotlist items with more familiar names, and you can add Internet sites directly to a Hotlist while off-line by typing in URL (Uniform Request Locator) names (from a book or magazine article, for example). The menu editor interface also lets you nest Hotlist items, so you can go around the Hotlist size problems with the Alpha 2 release, and also group Hotlist items by categories. Like this:

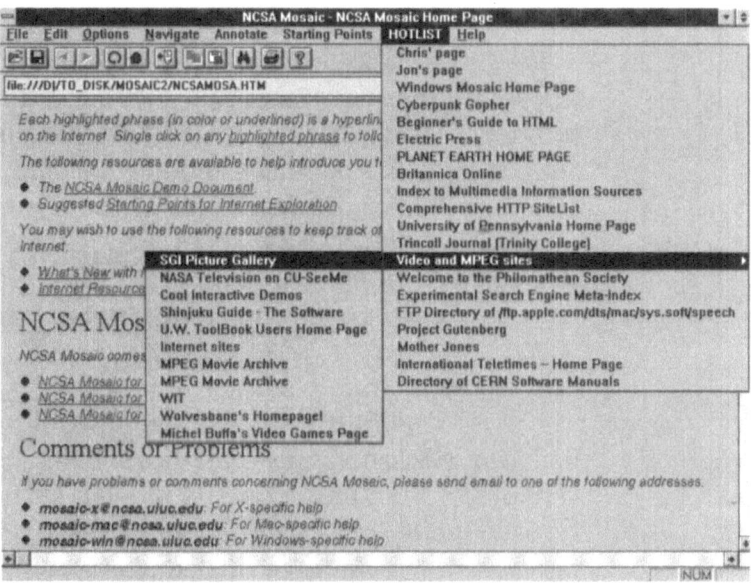

FIGURE 2.5.
Nested Hotlist under NCSA Mosaic Alpha 7.

The menu editor also allows you to edit the Starting Points menu, but that's only recommended for experienced users. The Starting Points menu is a useful preconfigured tool for cruising the Internet, and mucking it up could leave you without this resource.

You can also use Mosaic to run your file system directly, turning it into a type of file manager. You'll have a view of all of your files, and Mosaic will load text files directly into its main window (and use external viewers for all other data types). To try this, type `file:///d|/` in the Open URL panel, and press enter. This will show you a Mosaic view of your local D: drive. This should work for all Mosaic Web browser versions, including Netscape.

Spyglass Enhanced NCSA Mosaic

Spyglass was formed to exploit the commercial potential of software developed at NCSA. At first, it worked on high-end visualization programs. When Mosaic was developed at NCSA, Spyglass became the official licensing arm for it, and began work on an enhanced version that would be available from other companies. These include O'Reilly & Associates and IBM. Spyglass Mosaic is only available from licensees like this, as a part of a network software package or included in a book.

The enhanced Windows Mosaic program from Spyglass is an interesting version. Stemming from the development of 32-bit Windows Mosaic developed at NCSA, it requires the same 32-bit subsystem to run under Windows 3.1, and runs natively under Windows NT, Windows For Workgroups, and Windows 95.

Its main window offers the same main document view as in the other versions, but there's no button bar, just forward and backward keys. The document title appears in the top bar, and the URL listing and globe icon appear with the forward/backward keys under the menu bar.

The File menu has the same options for loading a URL or a local file, and also adds one for opening a new window. In this version, you can have more than one HTML document available for viewing at the same time, like opening more than one word processing document in the same program, and you can switch between them easily. This means you can have more than one Mosaic document on the screen at the same time, for example, for a side-by-side comparison, without having to launch another copy of the main Mosaic program.

With Enhanced NCSA Mosaic, you can also select the text area of a Mosaic document, and copy and paste it into another program, like Notepad. The Edit menu has an option for selecting all of the text in a document, as well.

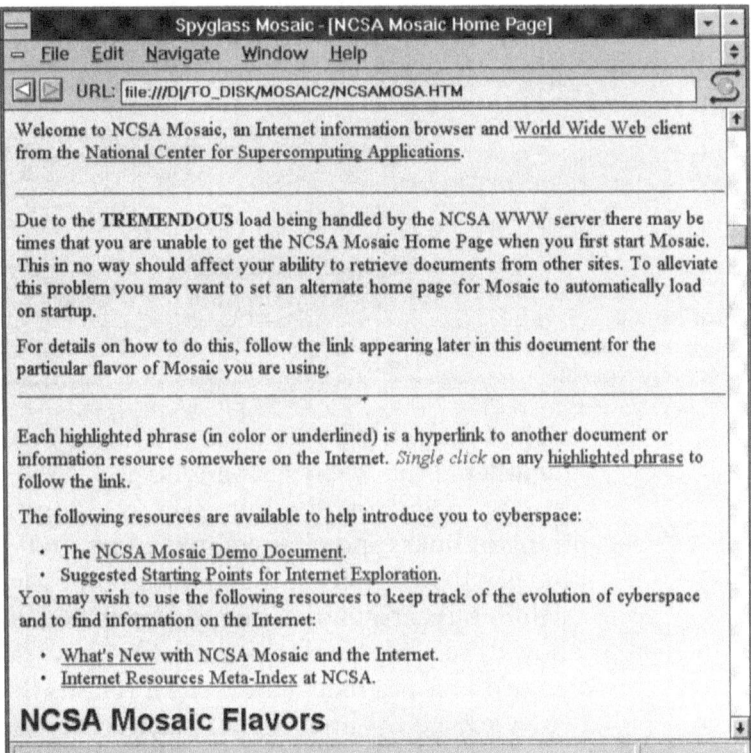

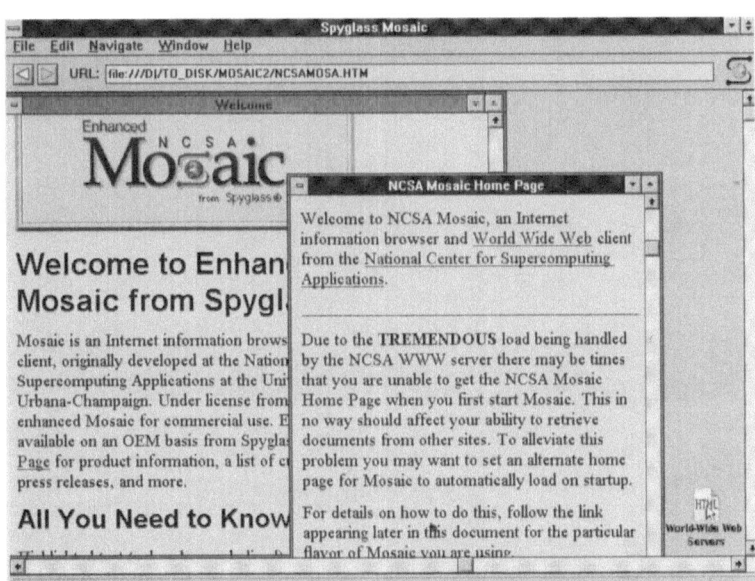

Also in the Edit menu are options for viewing a document's HTML source, and for customizing the main window appearance via a Preference panel.

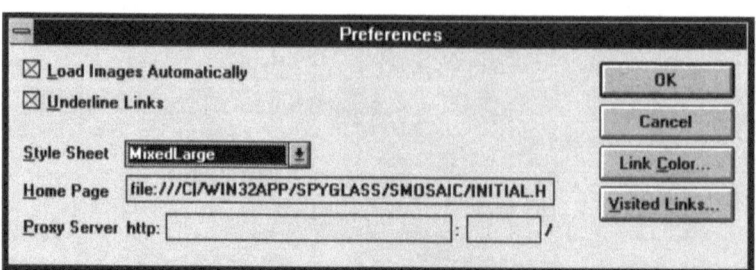

FIGURE 2.8.
Spyglass Mosaic's
Preference panel.

The Preference panel gives you limited control of fonts used in documents (between serif and sans serif, or a mix of both). You can also delay loading inline images, turn off automatic underlining of links, and change link colors. In this version of Mosaic, this is where you'll set the default home page as well.

Under the Navigate menu, you'll find a History box which shows links previously loaded (not just items loaded into the cache). This is a nice feature, as it retains the link addresses even when you go off-line. It also has a feature for exporting the history list as an HTML document, which can be used as a local home page.

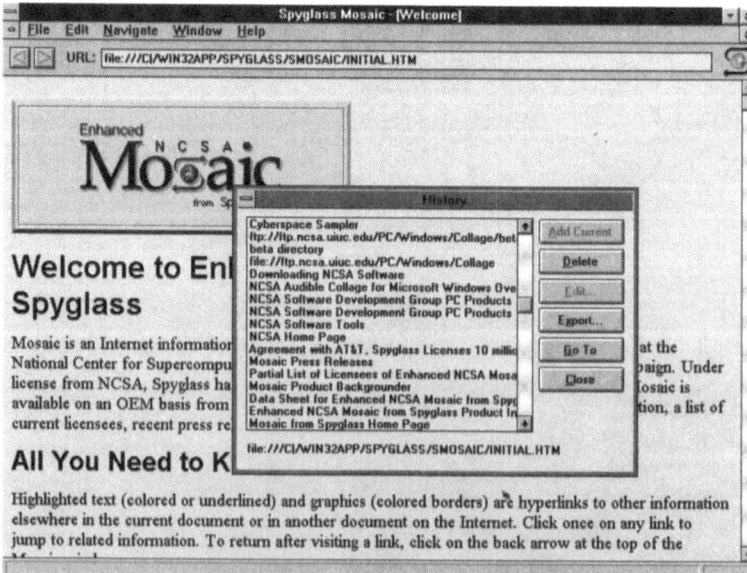

FIGURE 2.9.
Spyglass Mosaic's
History panel.

You can also do this with the Hotlist. Spyglass Mosaic's Hotlist appears in a floating window with its own scroll bar, and avoids the problems with NCSA Mosaic's menu version.

The Edit menu also has an option that allows you to load delayed in-line images, which means you can set the image delay function, and only load those images you choose, a nice touch.

The Help system for Enhanced Spyglass Mosaic is contained as a set of HTML documents. This provides a good description of the interface. Under the Help About menu, there's a technical support panel which gives diagnostic information on your system, including information on your Mosaic and Windows version, video and network setup, and printer configuration.

This version of Mosaic also adds a progress bar to the right of the status read-out line at the bottom of the main window. This is an improvement over the spinning globe icon, as the bar shows the relative progress of the network transfer, so you can visually judge how long a transfer should take.

As we mentioned previously, the 2.0 version of Spyglass Mosaic will add internal support for Adobe Acrobat files. Spyglass Mosaic can work with all of the external viewers available for other versions of Mosaic, as well.

AIR Mosaic from SPRY

SPRY's version of Mosaic for Windows, AIR Mosaic, was developed by one of the original designers of the NCSA version. As such it incorporates a lot of the good design that went into that product, and also adds some interesting innovations. It's available commercially as a part of Internet in a Box, the AIR Series network package, or in an FTP demo version (from `ftp.spry.com`).

The Main window features a full set of buttons for program control, and these can be set to display either pictures or text or both. The buttons are also laid out a bit better, and are easier to use. Below the button bar are two list items for Document Title and Document URL; these not only read out the current items loaded, but are also part of a drop-down history list. You can move to places you've visited, or documents loaded in the program cache, by scrolling down through either list. This is useful, as it also retains a history of your previous Mosaic session even after you exit the program, so that when you start up again, you have the previous session's list of documents and places available to browse through.

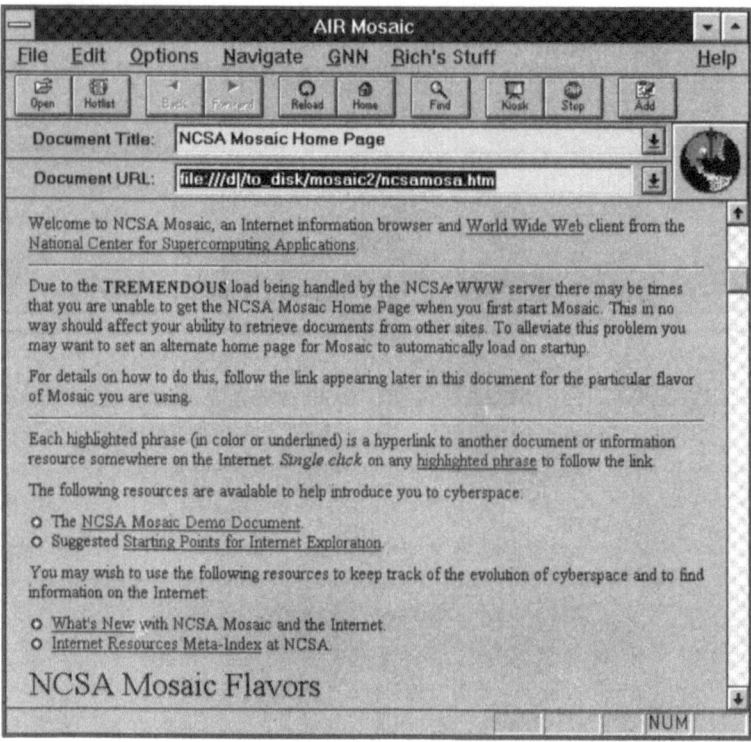

FIGURE 2.10.
AIR Mosaic's main window view.

To the right of those listings is the progress indicator, in this version changed from a spinning globe to a radar sweep. This icon also changes size when you show/hide the button bar and document title/URL fields, and you can stop it from being animated if you choose (these features are found under the main menu on the Options/Configuration panel).

The bottom of the window has a standard status line to the left, and features a larger set of standard Windows indicators to the right, but no progress bar.

The File menu options are the same as in NCSA Mosaic, with the addition of a Hotlist menu item. This Hotlist version is the best available, as it uses its own floating window to store Hotlist information in several nested files, and allows you to start new Hotlists, and save documents under these. You can also put any of these Hotlists on the File Menu for easy access. SPRY provides an extensive preconfigured Hotlist with **AIR Mosaic** in the Internet in a Box package.

The Edit menu is standard. There's no option to select text in a Mosaic document as with the Spyglass or Netscape browsers, a drawback. The Find panel is standard.

Under the Options menu, AIR Mosaic begins to differentiate itself from the rest of the pack. Besides the standard Load To Disk item, there's also one for Kiosk mode. This puts AIR Mosaic into a document window-only mode, with menus and buttons hidden, so that navigating can only be performed through the Hypertext documents themselves. This is useful for children or users with little computer experience.

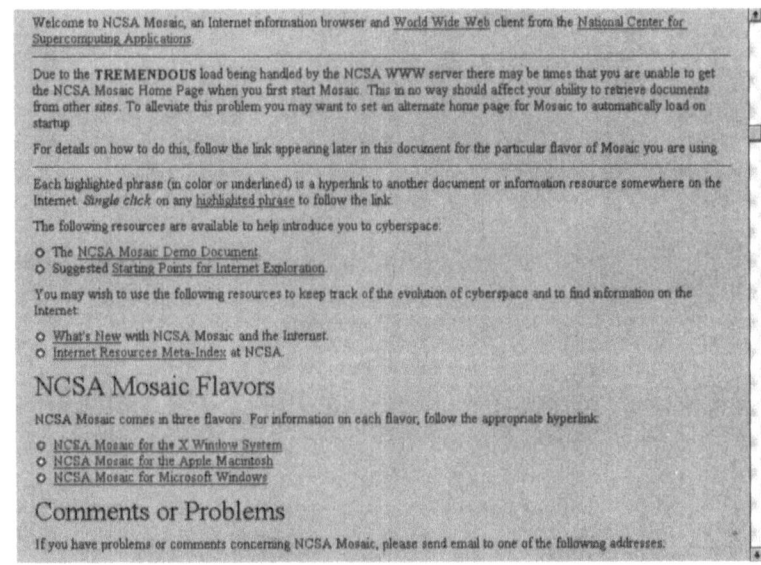

FIGURE 2.11.
AIR Mosaic in Kiosk mode.

The Configuration option panel opens a whole subset program under AIR Mosaic for configuring cache size, setting fonts and link colors, telling the program your email address and Network News server, and more. This facility is also used to configure viewers. A button at the bottom of the Configuration panel opens an interactive External viewer configuration panel, where you can add viewers directly or browse through your file system to locate them. A number of viewers are already listed, and some of these are preconfigured to Windows applications like Write and Multimedia Player; the others have to be configured later as you add viewer programs. This panel also offers you an easy way to define new viewer types and link viewers to them; this gives AIR Mosaic an edge over programs like NCSA Mosaic, where you'll have to edit the `Mosaic.ini` file to do this.

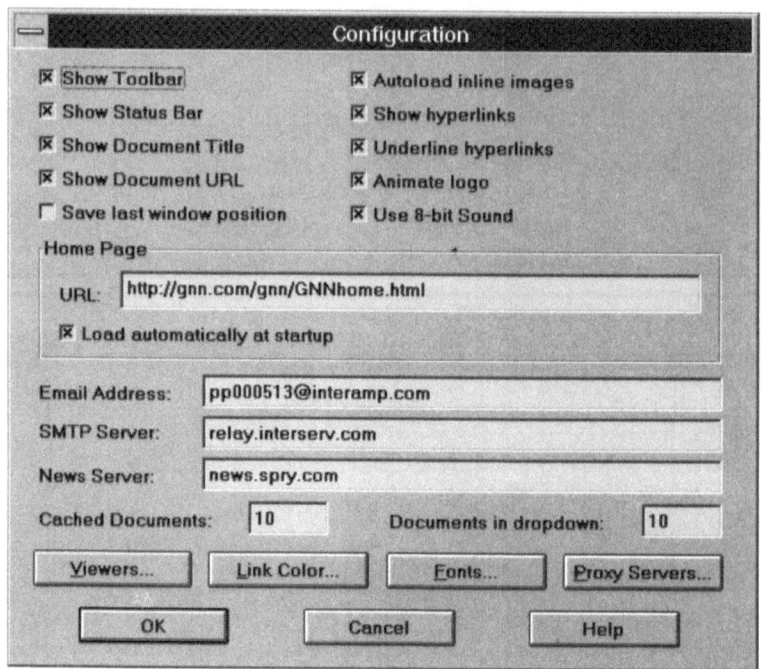

FIGURE 2.12.
AIR Mosaic's
Configuration panel.

Another button on the Configuration panel is used to set the fonts that AIR Mosaic uses. In addition to changing the fonts in several areas by using a standard dialog box, you can also use a simple enlarge/reduce button to increase the overall font size to a view you're more comfortable with.

The Options menu also has a nifty feature that allows you to import NCSA Mosaic Hotlist menus into AIR Mosaic. This translates the Hotlist file from the NCSA `Mosaic.ini` file into the interactive format that AIR Mosaic uses, and means that you can move up from the public-domain version to the commercial SPRY version of Mosaic without losing your existing Hotlist files.

The Navigate menu in AIR Mosaic is less well developed than in the NCSA versions, and isn't used for much more than moving between documents, accessing the History list, and adding documents to the current Hotlist.

AIR Mosaic also incorporates drag-and-drop, one of its more

interesting features. This means that you can drag HTML documents, text, and graphics files into the AIR Mosaic main window, and they will load directly. You can do this from the File Manager with local files, and with remote sites using an FTP program that supports drag-and-drop (like SPRY's Network File Manager, included with Internet in a Box).

Netscape for Windows

Netscape is one of the most exciting innovations to come about since the original development of Mosaic at NCSA. The team led by Marc Andreeson (formerly at NCSA), joined by Jim Clark (formerly at Silicon Graphics), have developed what some have termed the second generation of Mosaic. Generally, the program lives up to its claims, but it also fails to deliver some of the functions of the other versions of Windows Mosaic. Netscape for Windows is available from Netscape Communications via FTP at the `ftp1.mcom.com` and `ftp2.mcom.com` sites in the `/Netscape/windows` directory, and will soon be available as a commercial product from retail software vendors.

The main Netscape window shows the same basic view as any version of Mosaic, and includes the same pull-down menus and buttons for common actions. Netscape also includes a smart Location panel below the button bar. You can forgo typing in the redundant part of most Internet addresses, because the program can figure them out by itself. For example, the standard Mosaic address for Microsoft's web site is `http://www.microsoft.com`. In Netscape, all you have to type is the `www.microsoft.com` part, and the program will take it from there.

Under the Location panel is a set of directory buttons. These are unique to Netscape, and are directly linked to Netscape Communication's home page on the Internet. For example, the Guided Tour button will load a tour of the Netscape interface. Since the document is located at Netscape's site, it can be very up-to-date. There are also directory buttons for finding out what's new with Netscape, reviewing commonly asked questions, using Internet search tools and directories, and reading Network News.

Also in Netscape's main window is a progress indicator (in the lower right corner), a bright red bar that fills out to show the

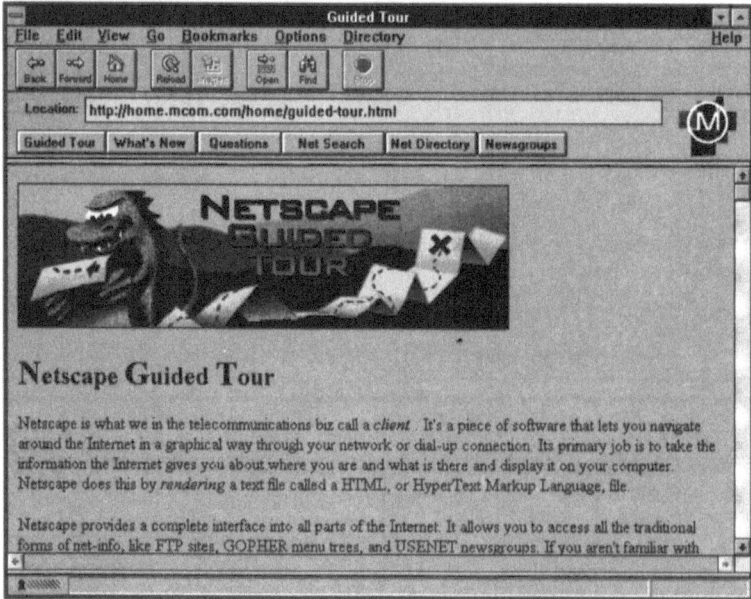

FIGURE 2.13.
Netscape's Guided
Tour view.

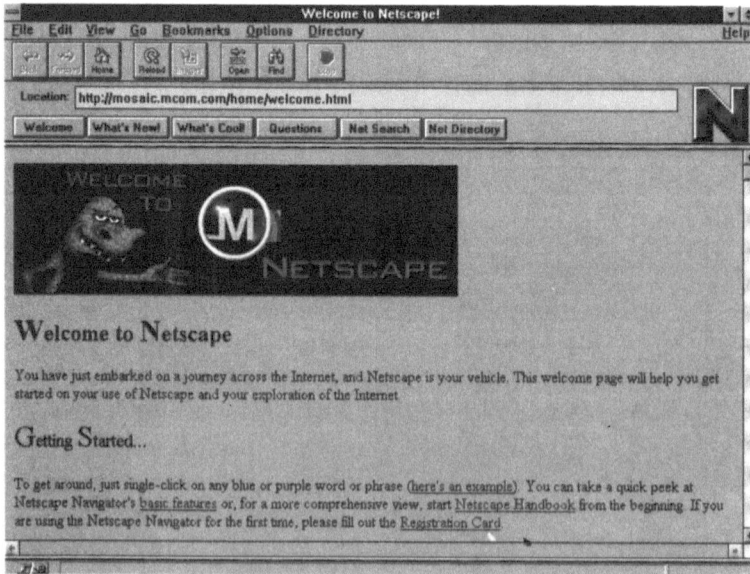

FIGURE 2.14.
Netscape main
window view.

progress of a transfer. This kind of tactile feedback is an integral part of using a Web browser, since it gives you a chance to judge the effectiveness of a transfer.

Netscape's File menu has standard items for loading URLs (known as Open Location in this version) and local files. You can also use the New Window function to launch another Netscape window, handy for looking at two Mosaic documents side by side, like with Spyglass Mosaic. The Mail Document feature can be used with a properly configured mail server to send the contents of a Mosaic page directly to another person via email. Print and Print Preview are standard.

The Edit menu in Netscape has a full range of cut-copy-paste functions, meaning that you can select and copy parts of the text in the main window, and transfer it to another document or file via the clipboard. The View menu has functions to load delayed images and view document source (HTML format). The Go menu has menu items to move back and forth between documents, and a pop-up History panel that shows a list of the Internet sites you've visited in your current session.

Netscape calls what it uses as a hotlist "bookmarks," and they work much the same way as in other versions. The Bookmark menu has a drop-down list as with NCSA Mosaic, and you can also use a separate Bookmark List box, like with AIR Mosaic. This box expands to become a full bookmark editor, with provisions for exporting and importing HTML hotlists from other versions of Mosaic, information on when a bookmark was created and last visited, and a panel for adding a description.

Under the Options menu, Netscape allows you to remove the toolbar, the directory buttons, and the location panel from the main view, to expand its range. The Preference panel is also located here. This is a single panel with a drop-down list for different sections, used to set page styles (including the default home page), configure directories and News feeds, set security levels, and configure helper applications.

Netscape is very extensible, and this is a good feature. It's also important to note that Netscape works interactively to configure itself during an Internet session. For example, if you attempt to load a file that Netscape can't use internally (like a picture file format such as .PCX), a warning panel will open up, and ask you

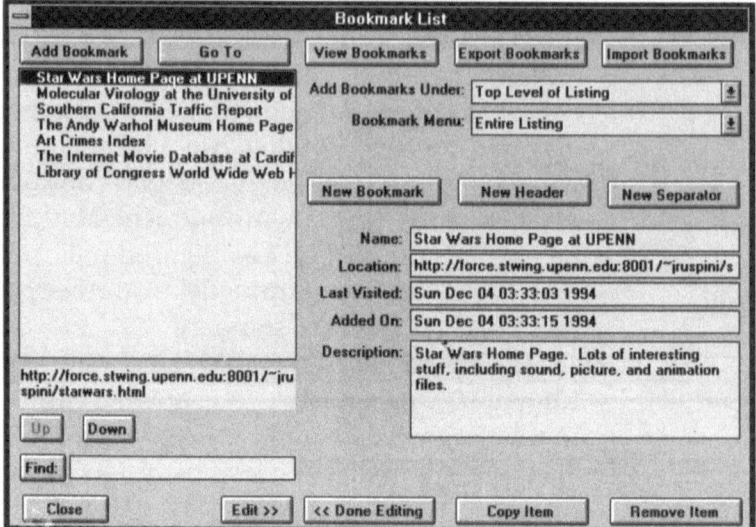

FIGURE 2.15.
Netscape's
Bookmark List view.

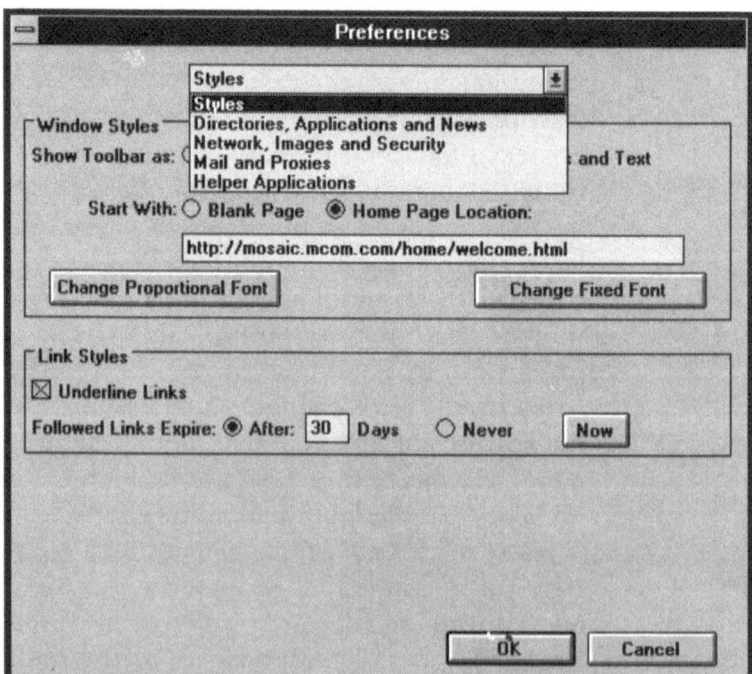

FIGURE 2.16.
Netscape's
Preferences panel.

if you want to load the file to disk, or configure a viewer for it. Once the viewer is configured, Netscape will remember it and use it the next time it encounters this type of file. The main drawback to this is that there doesn't seem to be a Load To Disk feature in Netscape; this mechanism handles all transactions, and won't let you load to disk a file it can recognize internally, unless you bypass it by shift-clicking on the link. You can save files to disk once they've loaded internally, however.

Netscape's Directory and Help menus are also linked to their Internet site. The Directory menu is related to the Directory buttons, and also includes a link to a "Cool Internet Sites" page. The Help menu includes version information and a link to a friendly on-line User's manual.

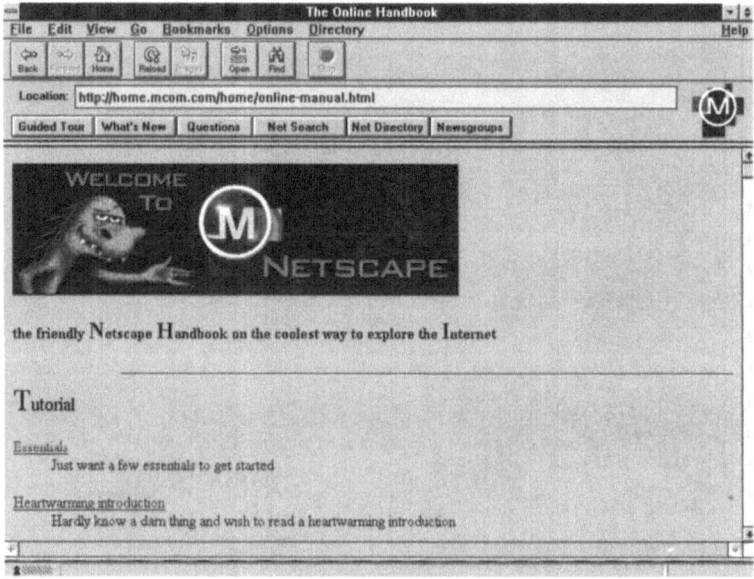

FIGURE 2.17.
Netscape's On-line Handbook.

WinWeb

WinWeb, from Enterprise Integration's EINet Galaxy, is an alternate Mosaic-like Web browser for Windows. It's also linked closely to an Internet site, like Netscape. The interface is less crowded than most browsers, and WinWeb doesn't have a lot of complicated features. You can find it via FTP at the `ftp.einet.net` site in the `/einet/PC/winweb` directory.

The WinWeb main window has a small row of buttons for navigating pages, launching the hotlist panel, and returning to the home page. Below these are status lines showing the current document name and URL. You'll also notice that the main window view of the EINet WinWeb home page has its own menu bar, with links to on-line help and search tools directly accessed from the page.

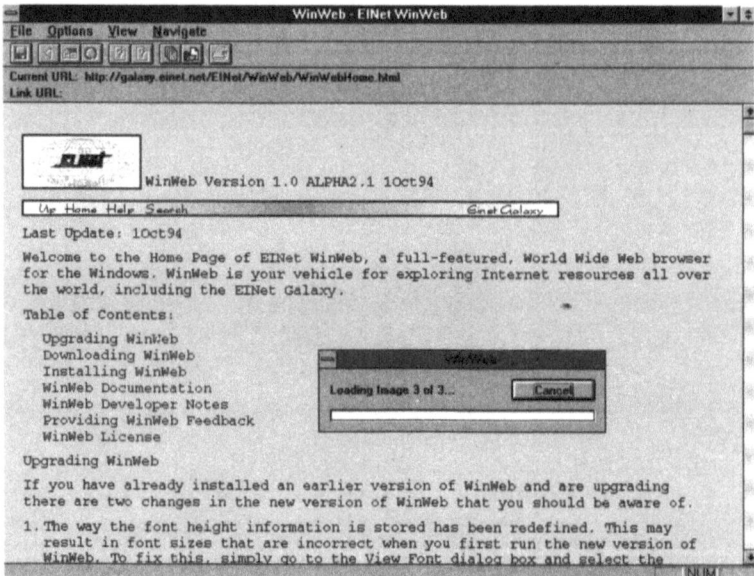

FIGURE 2.18.
WinWeb's Internet
Home Page.

The File menu has features to open local files and to print documents. The Options menu has a selection to set WinWeb to load files to disk (called "Load To File"), and a panel for changing the default home page. The View menu is an easy way to change fonts, highlight colors, and the default gray background. The Navigate menu is used to load URLs (not the File Menu), and it's also used to search the EINet Galaxy, an Internet index linked to many sites.

WinWeb also provides a floating progress indicator bar, like the ones used in Netscape and Spyglass Mosaic.

WebSurfer

WebSurfer is a new release from Netmanage, the makers of Internet Chameleon. Chameleon is a package of TCP/IP (Internet protocol) applications that connect you to the Internet. In this case, we use Netmanage Chameleon to connect to our Internet access provider, PSI. PSI provides us with the Internet Chameleon software, which now includes WebSurfer. You can also buy Internet Chameleon with WebSurfer directly, as it has an Instant Internet feature that allows you to configure it to popular Internet access providers with ease. It's available from retail software outlets, usually in the same places you'll find Internet in a Box.

The main WebSurfer window has a standard Web browser look. The File menus, button bar, and URL list box are much the same as in other Web browsers. There's a button to launch the Open URL panel, however, which is not found in other browsers, and it can come in handy. There's no progress bar, and instead of a globe icon at the upper right corner, there's a kaleidoscopic chameleon icon at the upper left.

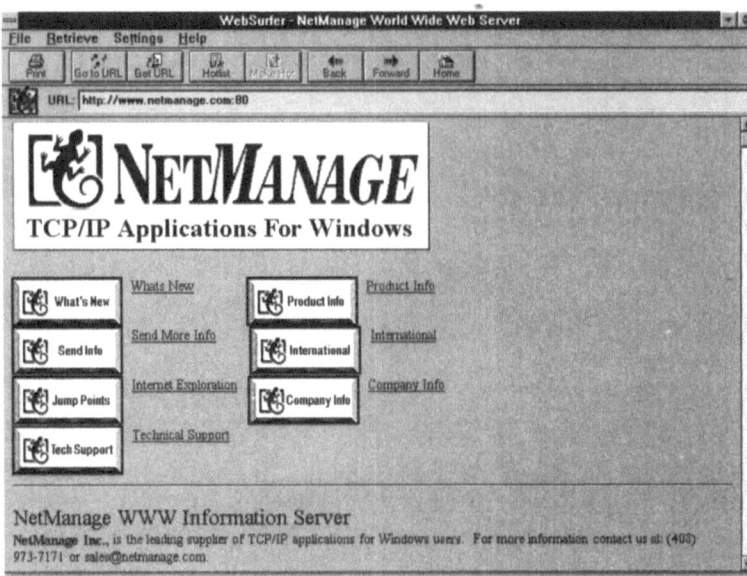

FIGURE 2.19.
Netmanage's
WebSurfer main
window.

The File menu is used to load, change, and save configuration parameters for WebSurfer, to print documents, and also to import Mosaic.ini hotlist information. This feature is basically an HTML converter for previously saved hotlist information in NCSA Mosaic, and makes it easy to move from NCSA to WebSurfer. You can also use the converted HTML hotlist with Netscape.

The Retrieve menu is where you'll find the Go to URL item (or use the Toolbar button), and the pop-up History and Hotlist panels. You also use it to launch the Properties panel, the HTML editor, and the Connection status screen.

The Properties panel is a document configuration screen for WebSurfer. It allows you to set parameters for a Web document, including changing the title and location (Internet address). It also shows the last time the document was accessed, and provides a panel for adding a personal memo.

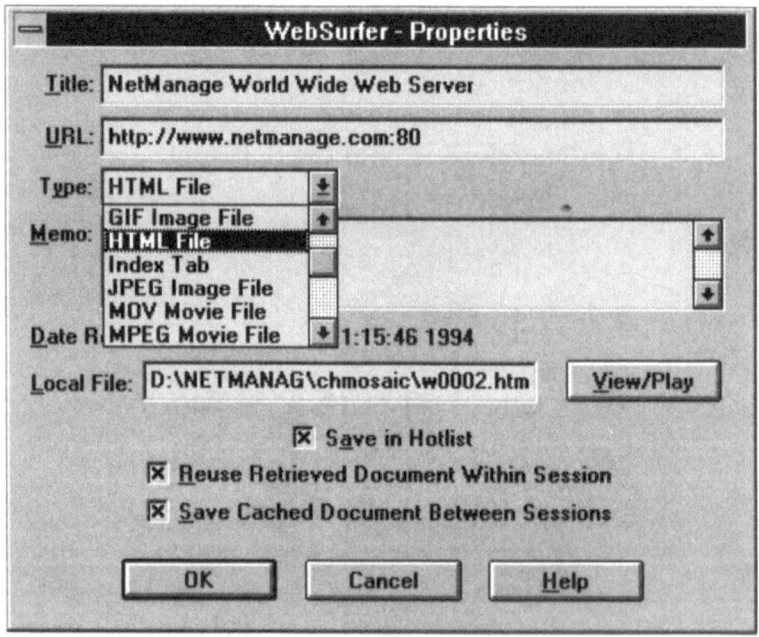

FIGURE 2.20.
WebSurfer
Document
Properties panel.

The default HTML editor for WebSurfer is Windows Notepad, but you'll want to change that (in the Settings/Preferences panel). Notepad won't wrap text properly, and your HTML document won't look right. To its credit, the HTML edit function is a better way to view a document's source file than in other Mosaic Web browsers, since you can load the HTML text for any file directly into a word processor or advanced HTML editing program.

The Connection Status panel shows the progress of your Web document retrieval operation, in terms of time, site names, and amount of information transferred (in bytes). It's also the only way you can stop a document transfer under WebSurfer, which doesn't feature a button or an icon you can click on to stop a transfer.

Under the Settings menu, there are further configuration options. The Preferences panel allows you to set the default home page, and to specify some helper applications (like the HTML editor), but not many. You can also set the image loading to be deferred, if you have a slow link.

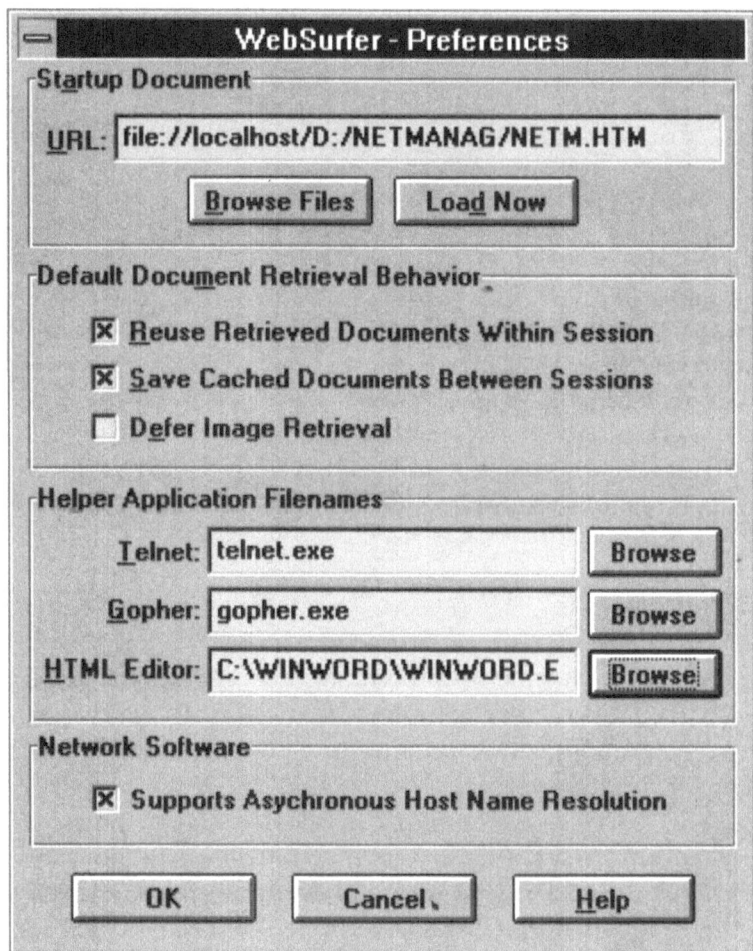

FIGURE 2.21.
WebSurfer's
Preferences panel.

The Style Schemes panel, also under the Settings menu, is a good graphical way to set various global attributes for Mosaic Web documents. It consists of two parts, one for Element Attributes (mostly font choices for various HTML headings), and Document Attributes (for changing document spacing, color, and overall font size). This panel makes it easy to customize Web-Surfer to your own preferred viewing options.

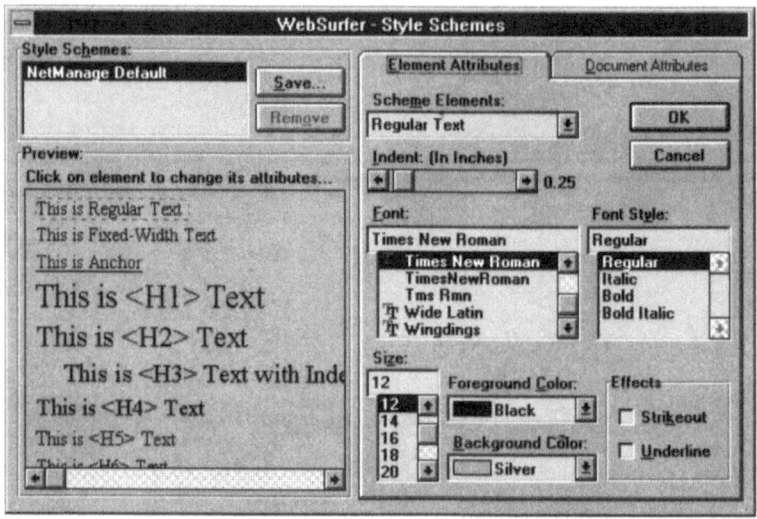

FIGURE 2.22.
WebSurfer Style
Schemes panel.

You can also turn off the toolbar and the dialog bar (the URL listing line). WebSurfer also allows you to turn off the bottom status bar, and gain more viewing space in the main window.

Viewers

Mosaic Web browsers for Windows can display a variety of file types in its main window, but not everything. Some graphics files and all sound files must be accessed via external programs called viewers.

Most versions of Mosaic can display .GIF files as in-line images; this means that they are available within the program. An external viewer has to be used to see .JPG compressed images (except for Netscape, which can display these internally) and PostScript, among other files not supported directly. The viewer is also used to access linked images that aren't displayed (for example, those in an FTP directory), and to access larger versions of in-line thumbnail-sized images.

Another image format you'll encounter when using Mosaic is the X Bitmap (usually .XBM); these are used by most Unix systems, and therefore are occasionally encountered on the Internet. The default image viewers LView and "Image Viewer" for AIR Mosaic don't support these types of images, but most Web browsers can handle them internally, including Netscape and NCSA Mosaic.

Spyglass has also announced that they will support Adobe Acrobat .PDF files internally, bringing this cross-platform document standard to Mosaic. Acrobat allows for documents created in separate word-processing and graphics programs to be displayed on other systems with their fonts and graphics intact, and it will be a welcome addition to text and graphics files available on the Internet. Versions of Mosaic without the ability to view Acrobat files internally will be able to call the Acrobat Reader as a viewer program.

How to Add Viewers

Adding viewers to Mosaic for Windows varies from the complexity of NCSA Mosaic, where you'll have to edit the Mosaic.ini file, to the simplicity of AIR Mosaic, where you can pick the viewers out of the Windows file system using a browser. Netscape works more or less the same way, and also prompts you to add a viewer if it encounters a file type it doesn't recognize.

Here's how to add a viewer in NCSA Mosaic; this will also work for the Spyglass version (where the file to edit is called smosaic.ini).

First, open the Mosaic.ini in Notepad and locate the section of the file called [Viewers].

Notice that several file types are already listed, in the format data type/data name. TYPE0="audio/wav" means that the audio format .WAV will be recognized by Mosaic (this is a standard Windows sound file format). Farther down the list, the audio/wav="mplayer %ls" line means that when Mosaic encounters a .WAV file, the Windows application Multimedia Player (mplayer.exe) will be launched, and the file will be loaded. This is already configured, and as long as your Windows directory is installed in your `autoexec.bat` file path statement, this viewer will work.

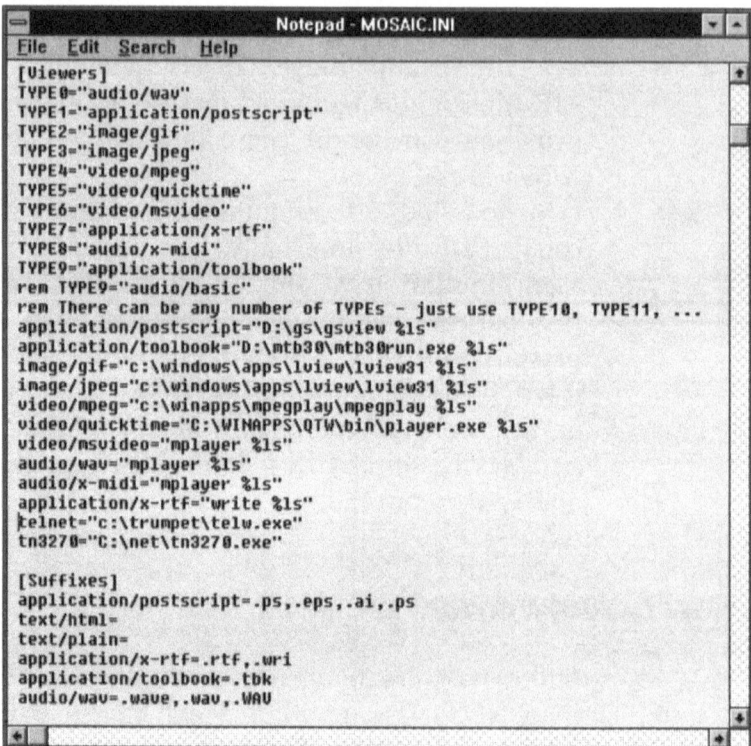

FIGURE 2.23.
`Mosaic.ini` Viewers and Suffixes sections.

The [Suffixes] section below the [Viewers] section is used to tell Mosaic what the file extensions it encounters means.

What's not supported directly in Windows for external viewing are the main graphics types .GIF and .JPG. To install the viewer for these, use Mosaic to access the Windows Mosaic home page, and go to the link marked "Viewers." Set Mosaic to Load To Disk, and click on the LView item. The LView ZIP file will be transferred to your hard disk.

Edit the `Mosaic.ini` file (or use the shareware Set Mosaic graphical program) to match the directory where you put LView after you install it, and you'll have an external viewer ready for Internet graphics.

Follow these procedures with the other viewers listed, and you'll soon have a full set of Mosaic helper programs, capable of

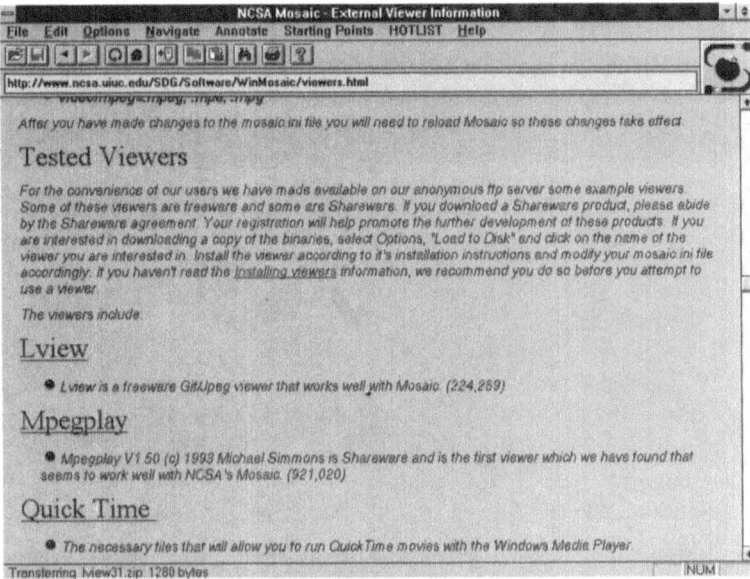

FIGURE 2.24.
Downloading LView from Windows Mosaic's Viewers page.

viewing graphics and playing a wide array of sound files and animations. Remember, you're not limited to the players listed at the NCSA site (you may want to use an alternate MPEG player, and to use Apple's QuickTime for Windows, for example), and you can use whatever external programs you want. LView (also known as ImageView for Internet in a Box) is a good fast player, though, and is recommended for use with Windows Web browsers. It also supports drag-and-drop, so you can use it to view files from the File Manager just by dragging them into its main window.

Netscape allows you to configure viewers on the fly. After you install a viewer, you can set it up for Netscape by attempting to access a data type for your viewer. Netscape won't recognize the file (it hasn't been set up yet for it), but it will ask you if you want to configure a viewer, and allow you to browse through viewer types.

For example, if you install an MPEG player, you can then go to an MPEG archive with Netscape, and click on an MPEG movie link. Netscape will ask you if you want to configure a viewer, and you can then select your new MPEG player. Netscape will then be configured for all future MPEG files with that viewer.

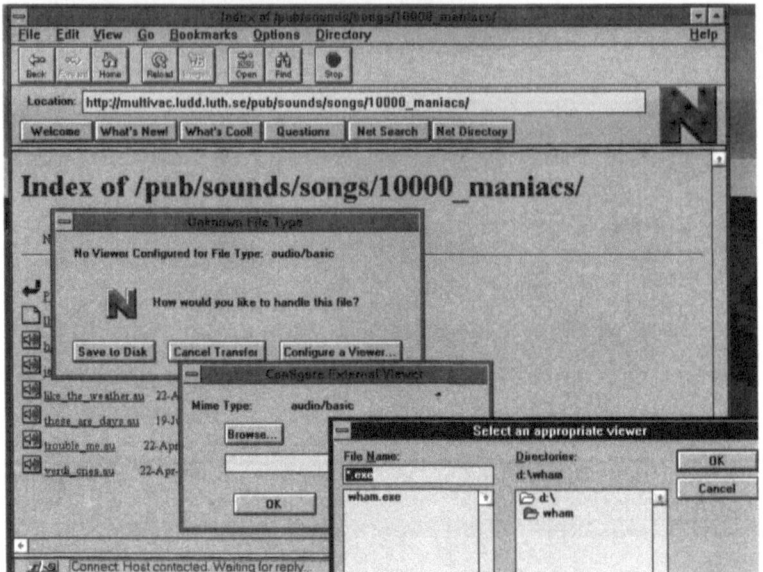

FIGURE 2.25.
Netscape's automatic viewer configuration panels.

3
The Access Provider

Choosing an access provider is the most important step you'll take in running Mosaic. There are a number of different types of service providers that give access to the Internet. First, we'll look at a shell account.

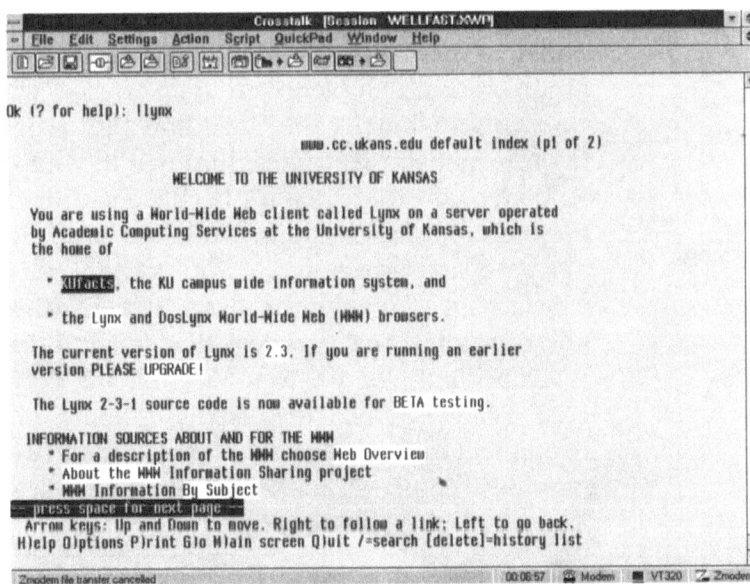

FIGURE 3.1.
A shell account on the WELL in Sausalito.

The shell account is an indirect connection to the Internet. This means that what you actually get when you connect to your service provider via modem is a session running on the remote system, not on your computer. Depending on the programs available on the remote system itself, this account may have a wide range of Internet capability, but it won't be able to run Mosaic.

This is because Mosaic is a client program. The information Mosaic uses needs to load directly into it as it runs on your system, not on a remote system.

A shell account uses a store and forward method of transaction. Your shell account usually gives you a personal file directory on the remote system. When you FTP a file from a remote site, it comes to your directory on the remote computer, not directly to your local system. You then have to download it from that system to yours.

Mosaic uses a method of transaction known as TCP/IP (Transfer Control Protocol/Internet Protocol) to pass information directly from computers on the Internet to itself. These protocols were originally designed for high-speed computer networks, but versions were subsequently developed that could run over standard telephone lines. SLIP (Serial Line Internet Protocol) and PPP (Point To Point Protocol) are terms for types of access to the Internet that standard modems and ISDN equipment can use to run graphical interface programs like Mosaic.

There's more to it than that, however. The shell type of account usually provides you with some kind of actual service organization (like the WELL, the Whole Earth 'Lectronic Link, a collective of minds that will be able to help you get your account running smoothly), whereas the SLIP and PPP types of accounts bring the Internet connection right down to your own system, and often leave you there. That's because the shell account runs on a remote computer system, while the SLIP/PPP provider is more of a gateway right down to your system.

Selecting an Internet access provider with a proper level of support is crucial. An example provider is PSI (Performance Systems International), of Reston, Virginia (at `http://www.interramp.com`). What PSI provides is the connection for your computer to connect to the Internet. As part of a start-up package, PSI also provides the necessary software (at an extra fee). That software is Netmanage, Inc.'s Internet Chameleon, a set of Windows tools for connecting to the Internet. Internet Chameleon now includes the WebSurfer Mosaic-style Web browser, mentioned earlier.

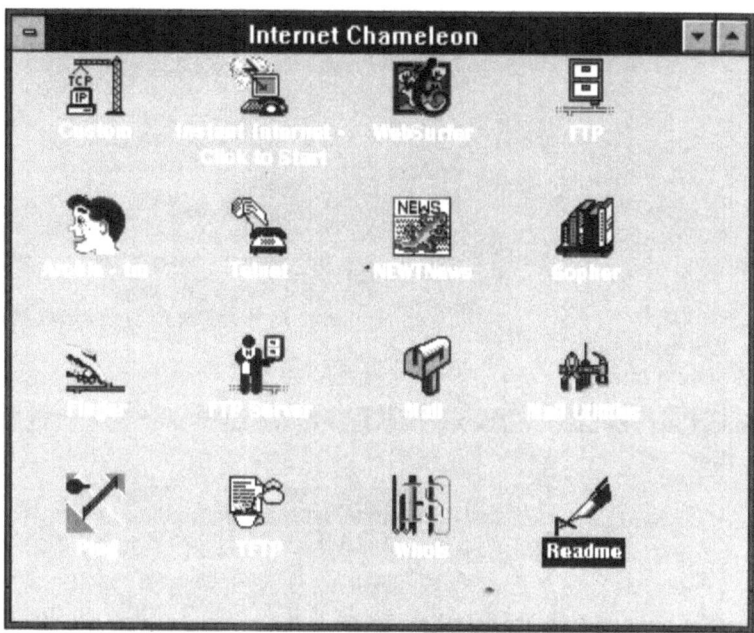

FIGURE 3.2.
Netmanage Chameleon program group.

PSI provides the telephone numbers and system addresses for your account. It's important to understand fee rates at this point. The Internet access provider often charges a flat rate per month for the account, and an hourly fee for as long as you're connected. PSI's initial rate was $29/month for up to 29 hours of connect time (about an hour per day), and $2/hour for over 29 hours per month. This assumes that you're near a POP (point-of-presence) set up by the provider; this is a regional phone number link to your Internet account, and means that you'll be making a local call to establish your SLIP or PPP link. Otherwise, you'll be adding a long-distance phone bill charge to your Internet account cost. Make sure that your Internet access provider has a local call access point.

The Netmanage application Custom is used to connect to the PSI InterRamp service. It's a cross between a modem program and a network interface. Custom launches Newt, a Windows Socket-compliant interface. The Windows Socket standard is a networking protocol that allows Windows programs (like

Mosaic) to receive networked information. Establishing a valid WinSock connection is crucial. The Custom interface includes a dialer and configuration screens for setting up modems, serial ports, and host servers.

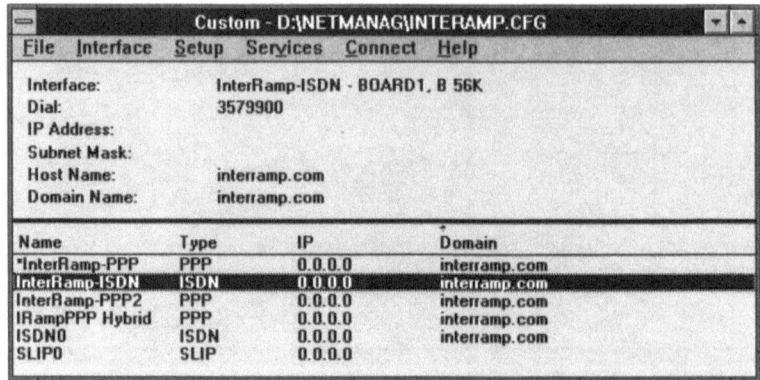

FIGURE 3.3.
Netmanage's
Custom Internet
connection program.

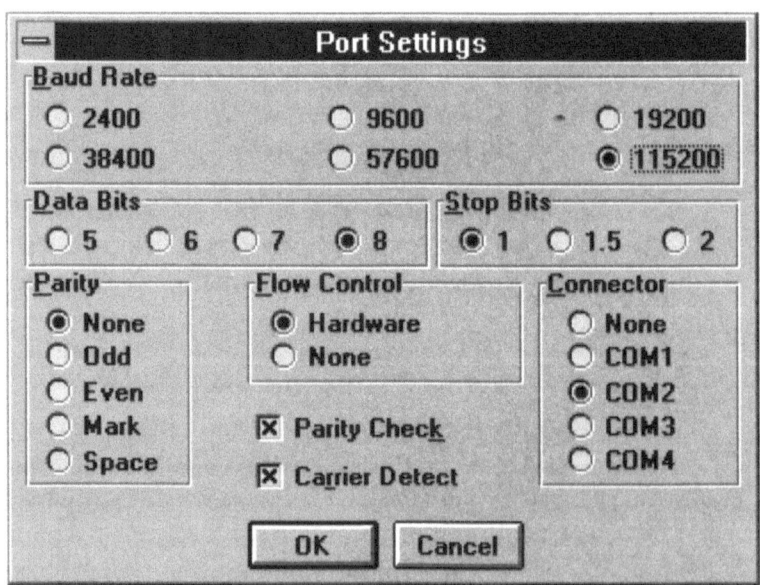

FIGURE 3.4.
Custom's port
settings panel.

The host server is where your Internet account will connect. Mosaic needs a properly configured host so that it can find Internet places by name. The standard address form for an Internet site is a four-segment dotted decimal number (for example, 192.9.201.255); this corresponds to a name (i.e., www.foo.com).

A Domain Name Server is used to translate names into the numerical addresses computers use.

The Custom application connects to the Internet by dialing out via your modem, launching the Windows Socket interface, and establishing a connection to the Internet server. Once you're properly logged in (at this point, your access provider will probably require you to use a password to access your account), your Windows applications can use Internet services, and you can use Mosaic.

What you might find, however, is that your Internet access provider's software doesn't include Mosaic, or that you'd prefer to use another version. You can use Internet access tools like FTP to download an alternate Web browser. For example, we could use Internet Chameleon's FTP application to go directly to the NCSA site for original Windows Mosaic, or to Netscape Communications archives for the latest version of Netscape for Windows. FTP stands for File Transfer Protocol, and it's used to move files from a remote site to your local system. Mosaic can use it directly, but you have to have Mosaic first. You can also find shareware FTP programs with good Windows interfaces, like WS_FTP, designed to run as a Windows Socket compliant application. FTP programs for Windows all have a fairly similar interface, and can locate directories in the same manner as Windows' File Manager.

Once you connect to a server like the one at NCSA, move to the relevant Windows directory, and locate the Mosaic program you want to use (either a 16 bit or a 32-bit version). The files will be in a compressed binary .ZIP format, so make sure that your FTP program is set to transfer binary files. Download the file by highlighting its name and hitting the copy button; the file will transfer to your local directory. If you're using Internet in a Box, you can use the Network File Manager in the same manner, and drag files from remote directories to your hard drive to transfer them.

Create a directory for the Mosaic file you retrieve, unzip it to that directory, read the instructions, do the necessary installation steps, and you're on your way to running your preferred version of Mosaic.

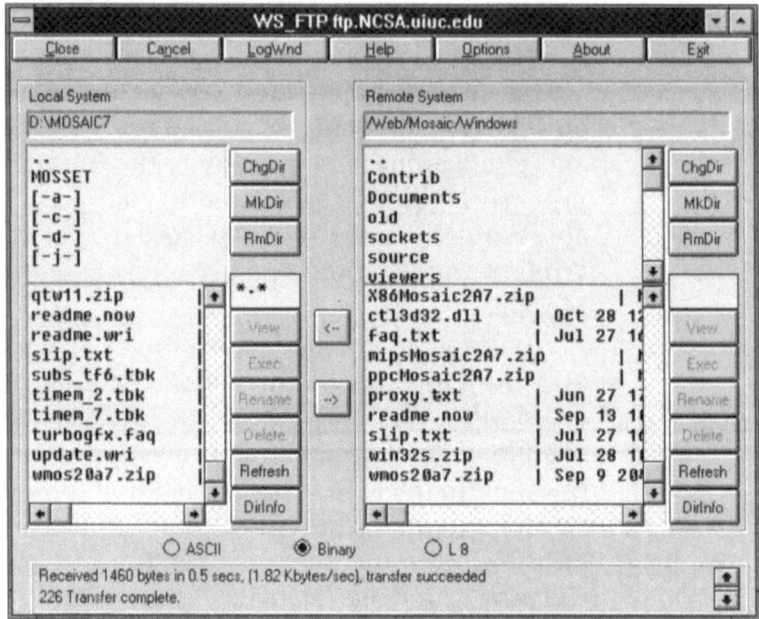

FIGURE 3.5.
WS_FTP accessing
the NCSA FTP site
for the latest version
of NCSA Mosaic.

Alternate Access Schemes

Internet in a Box, from O'Reilly and SPRY, offers an alternate way to connect to an Internet access provider. Their software is preconfigured to set up an account with a service provider called Nov'x Interserv. As a part of the installation process, the software will dial an 800 number, log in, find a local number (if available), and establish a SLIP-style connection.

Since Nov'x will then become your Internet access provider, it's important to decide if you can accept their access charges. At this time, the monthly rate is $8.95, with an hourly charge of $8.95; this is a lot more expensive than PSI, for example, at $8.95/month and $1/hour (for the first 29 hours). Comparative charges for 1 hour per day = 30 hours per month would then be approximately $39.00 per month for the PSI account and $278.00 per month for the Nov'x account. Caveat emptor.

It's fortunate that SPRY's Internet software is easily configurable for a different Internet access provider. Its Configuration Utility is much easier to use than Netmanage Chameleon's.

To set up an alternate access provider account with SPRY's Configuration Utility, all you have to do is run the alternate installation program or launch the application from the Internet

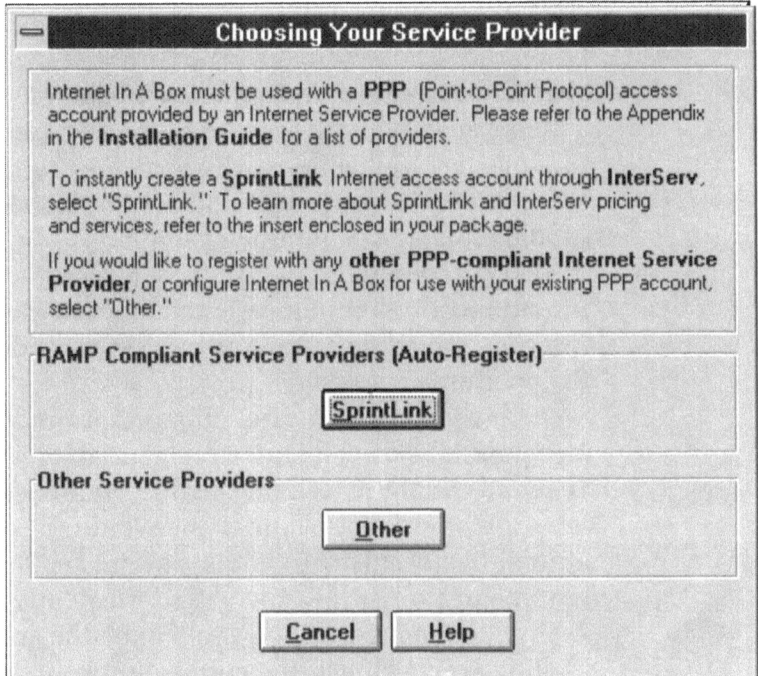

FIGURE 3.6.
Internet in a Box's
automatic PPP
configuration
screen.

in a Box program group. You'll see a set of buttons used to access configuration screens to register an account, manage the account setup, and launch alternate account profiles. The Auto Registration button will register your software with SPRY via a toll-free number and either set up an Interserv account or run an installation routine for an alternate provider. You can run the same installation routine manually by hitting the Manual Setup button.

The main configuration screens are as follows:

The Modem Settings window is where you'll set your serial port and modem type. SPRY has provided a long list of compatible modem configurations, so you won't have to fiddle around with modem command strings. Netmanage Chameleon's Custom application, by comparison, is configured for only three modem types.

Next follows the Dialer setup. This is the application that will make the connection to the Internet, so it's important to set it up properly. Enter the PPP access numbers from your service provider here, as well as your network settings (your Internet Protocol IP address, the name of your Internet host and domain, and your Internet name server). All of this information should be readily available from your Internet access provider, and is specific to whatever account you establish.

Remember, you have to have the account started first before the Internet applications will be able to log in and reach it. Only the Interserv account will register and connect directly without establishing an account beforehand.

After you configure the Dialer, you'll set the Default Hosts. This is used mostly to configure email, but you'll also use it to set your Internet applications' default servers and home pages, including the one for AIR Mosaic.

Once your software is configured properly, the AIR series Dialer should be directly linked to your applications. This Dial-On-Demand capability is a nice feature, and means that when Internet applications try to launch the Windows Socket, the AIR Dialer automatically starts up, dials, and makes the connection. What that means to AIR Mosaic is that it will automatically make the Internet connection when you launch the application, instead of making you establish the connection yourself, as with Netmanage Chameleon's Custom application. Note that this is a part of the underlying software, and doesn't mean that you won't be able to use any Internet access provider you want; PSI's PPP account works well with both the Netmanage software and with SPRY's AIR series.

Netmanage has risen to the challenge of providing an easier way to set up a Net connection by developing Instant Internet, a part of their latest version of Internet Chameleon. This provides a way to connect to several different Internet access providers (including PSI's InterRamp Service, IBM's Internet Connection, Alternet, Portal, and CERFnet).

The Instant Internet software allows you to register an account automatically, using a standard modem, by entering in your registration information and your credit card number. You can also elect to view the service agreement before you commit to a particular service. This provides in-depth details on charges (like connect-time fees) and local phone numbers. Use the More Info button on a particular service's main screen to access this information.

After you review the fine print, you can use the sign-up pro-

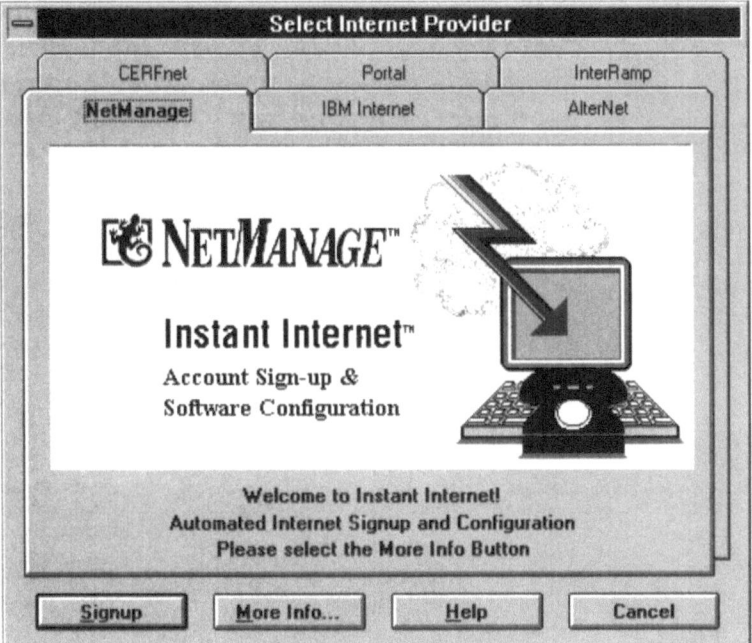

FIGURE 3.7.
The Instant Internet program, offering a choice of Internet providers.

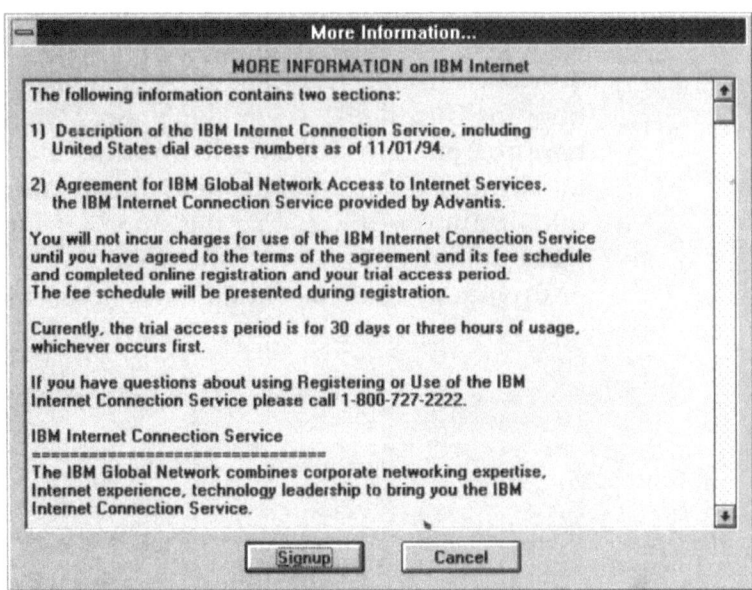

FIGURE 3.8.
IBM Internet Connection's information panel for Instant Internet.

gram to dial the relevant service. Sign-up will establish a connection (using an 800-number direct-dial connection, so your credit card information won't go over the Internet), register you with the service, and find you a local phone number for access. In the lower right corner, you can view the progress of your transaction through a series of connection icons.

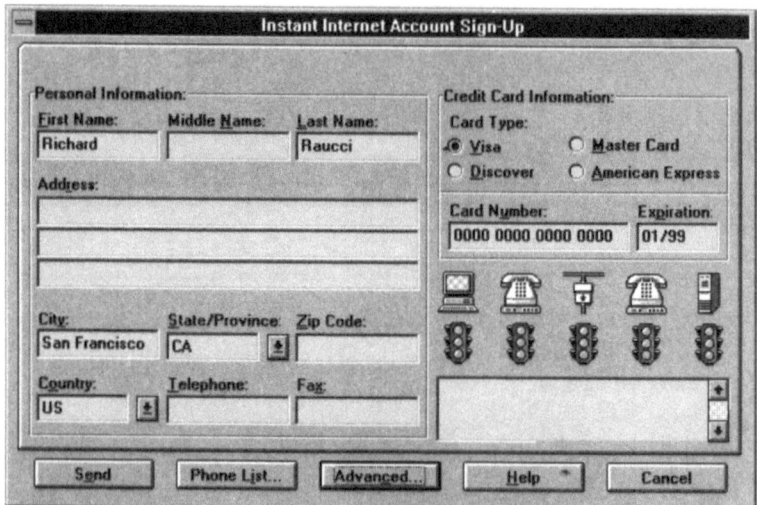

FIGURE 3.9.
Instant Internet
Sign-up panel.

Instant Internet offers a larger number of well-priced Internet access providers than Internet in a Box, but has a significant drawback in that it doesn't provide a list of modem configurations (as IBox does). If you don't have a standard modem, you'll have to configure it yourself from the sign-up panel. Click on the Advanced button, and you'll find a panel for adding modem initialization settings. Use the same settings as for your other telecommunications programs (like Procomm, Microphone II, or Crosstalk), and you should have no problems.

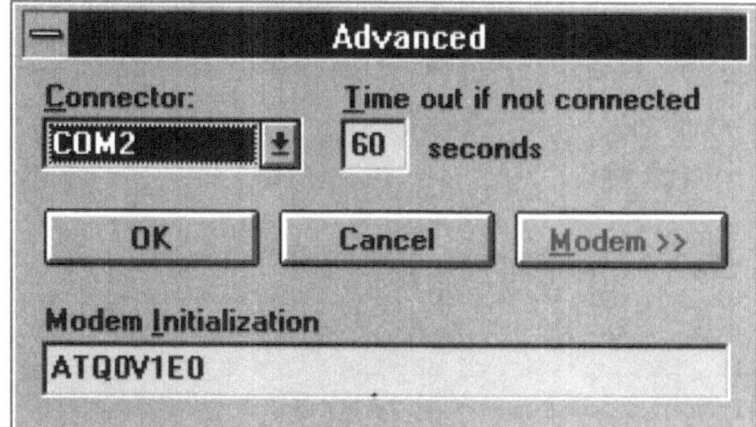

FIGURE 3.10.
Instant Internet's
Sign-up Program's
Advanced Modem
option.

Going It Alone

You can also configure a Windows Socket connection to the Internet using shareware. You still need an Internet access provider, but it doesn't have to be a commercial service. As long as your provider (for example, a university) gives you the necessary configuration information, you can use software like Trumpet WinSock to make your Internet connection (Figure 3.11).

Peter Tattum's Trumpet WinSock (version 2.0B) is a Windows Socket shareware application now available for SLIP and PPP connections. The software is available from FTP at many sites, including `ftp.trumpet.com.au` in the `/ftp/pub/winsock/` directory. You'll have to configure it yourself, including adding network parameters and dial-up script information (Figure 3.12).

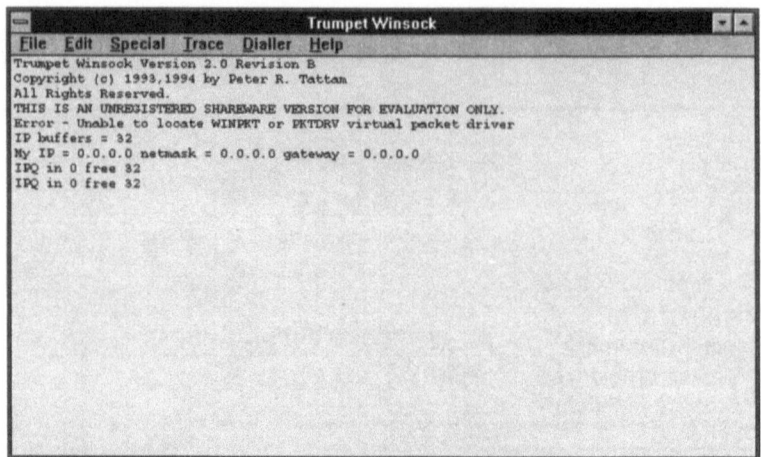

FIGURE 3.11.
Trumpet WinSock's main screen.

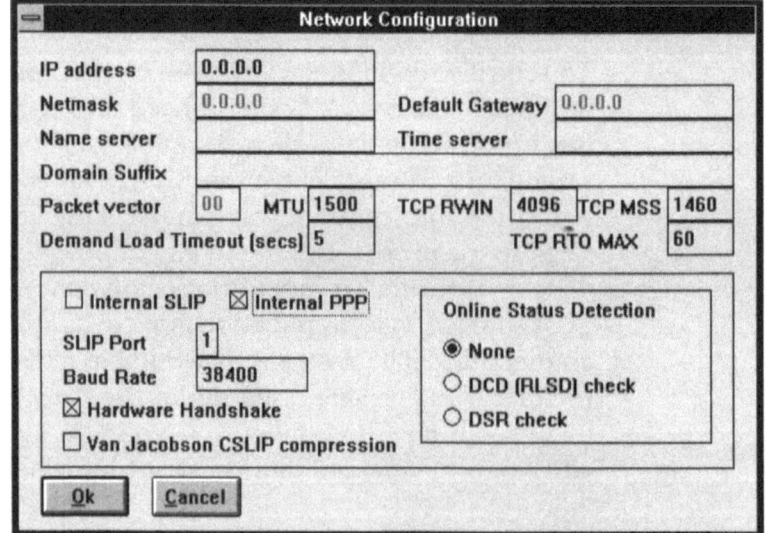

FIGURE 3.12.
Trumpet WinSock's Network Configuration screen.

Configuring Trumpet WinSock can be difficult, and is recommended only for users with experience in setting up dial-up network connections. Remember, the SLIP and PPP software sold by SPRY and Netmanage can also be configured to work with other Internet access providers, and they're easier to set up.

4
Modems and ISDN Adapters

Once you've got the PC configured and the access provider chosen, the next thing to do is to connect via modem. This chapter covers modems and modem protocols, and internal vs. external modems. It also discusses serial ports and UARTs (Universal Asynchronous Receiver/Transmitter, a chip that your serial port uses to transfer data) for ensuring trouble-free data communications. We then look at how ISDN is bringing even higher-speed digital connections to the desktop.

Choosing a modem is very important to running Mosaic over a telephone line. It's important to note that Mosaic needs the highest-speed connection you can get to your Internet access provider. The hardware you use is crucial for this. Mosaic transfers pictures, sound, and program files directly to your system, and it takes a lot of speed to keep up with the information flow. A minimal modem speed can slow your transfers down and significantly degrade your Mosaic experience. Get the fastest modem you can.

A 9600-baud modem is the bare minimum for running Mosaic, and most versions suggest a 14.4K-baud speed. While a text-based Internet service can run on a 2400-baud modem, it's too slow for Mosaic. Comparatively, a 14.4K-baud modem is six times faster at transferring data. Access providers and modem companies are also working at ways to increase throughput via special data-compression techniques. These features, found on the latest 14.4-baud modems, can increase their speeds to 19.2K, 28.8K, or even 57.6K baud.

The increased speeds are factors of the efficiency of your Internet connection software as well as your modem. For example, the Modem configuration screen for Netmanage Chameleon allows you to select a 19.2K connection even with a 14.4 baud modem, and SPRY's AIR Series will attempt to set the same modem up with a 57.6K link.

Differences in your Internet access provider now play a part in the overall connection. The trade-off lies in the best price/performance relative to speed. A 19.2K connection is adequate, but you should try for the highest connection you can get—57.6K if possible.

The path is now as follows: modem to Internet access software to Internet access provider. The connection from the PC to the modem is also very important, however.

A modem uses a UART to process data. This chip is built into internal modems and is a feature of serial port cards (and integrated serial ports) for external modems. The UARTs originally designed for serial port use on PCs were not very efficient at moving data, and these can cause problems for a program like Mosaic that needs to have fast, reliable data throughput. The 8560 UART is fine for a mouse, but bad for a modem. Next in line is the 16540 UART, also not the best for running Mosaic. The high-speed UART you need is the 16550. This double-buffered circuit is specially designed for high-speed transfers.

To check to see what UARTs are available on your system's serial ports, use the Microsoft System Diagnostics program (MSD.EXE). Exit Windows and run the program from DOS.

Note your specific UARTs. If you have 8450's for the first two COM ports, it's okay to leave these as is. Install a high-speed serial card set to COM 3 and 4 (make sure it has the 16550 UARTs) into the system. Make sure the system IRQ lines are separate. The IRQs are used for your system devices to pass data to your PC. Scan for clear IRQs using MSD.

An example conflict to avoid: a sound card set at IRQ 5 means that you can't set the second serial card's COM 3 or 4 to IRQ 5.

Also, locate the IRQs that the first two serial ports are using and make sure you don't try to set the second card's ports to these.

It's entirely possible that your newer system has high-speed serial ports; use these with the mouse at COM 1 (standard) and the modem at COM 2. Otherwise, add the second serial card, and connect the modem to COM 3 or 4. Make sure you don't have any other device conflicts before you add the second card. For example, one SVGA video card used the COM 4 line (not common, but it happens), and wouldn't work if another COM 4 was in the system. Disabling the COM 4 port left the COM 3 port available for high-speed modem use.

Back inside Windows, pay attention to the Ports settings in the Windows Control Panel. You'll want to set the COM ports to the proper IRQs and addresses for the new ports in order to have Windows recognize them. DON'T move the mouse from COM 1 or 2; Windows won't recognize it.

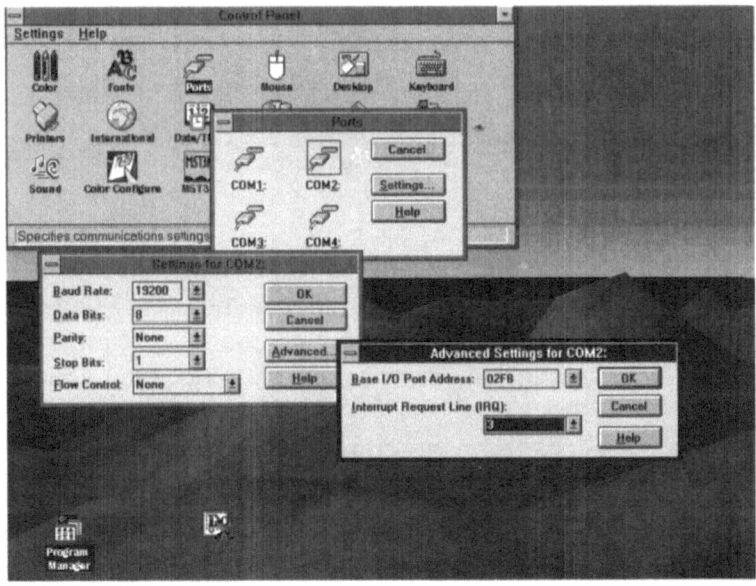

FIGURE 4.1.
Windows 3.1's Control Panel Serial Port configuration options.

You can also add a high-speed communications driver to your Windows installation. These programs are replacements for Window's communications driver (comm.drv), and allow better data throughput. An example program, KingCOM, also handles multiple COM port program conflicts. KingCOM effectively "tells"

Windows about the 16550 high-speed UART and allows Windows programs (like Mosaic) to access its higher functionality.

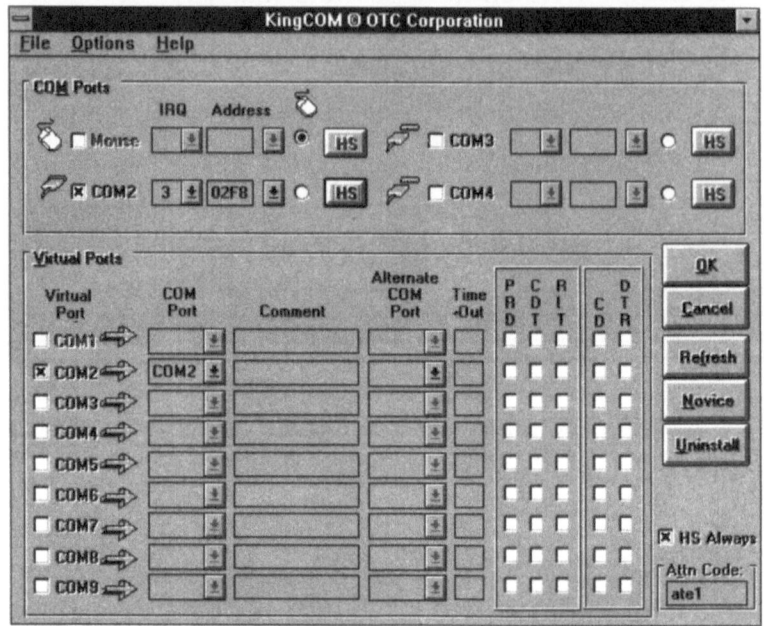

FIGURE 4.2.
KingCOM's main screen, showing configured high-speed serial ports.

Two modem examples for Mosaic for Windows use include Intel's low-cost 144EX external modem ($99) and Microcom's Deskporte FAST ES 28.8 (around $235). These are both readily available through computer retailers and mail order. The Intel unit comes in a small steel box and attaches to a standard serial port. You'll then need to configure your Internet access software to recognize your modem to be able to run PPP over your phone line.

Netmanage comes with a relatively low number of modem profiles (compared to SPRY's AIR Series, with over six hundred modems listed), and you'll have to adjust the default Hayes settings in the Modem configuration screen after selecting a PPP link in the Custom application.

Use the modem initialization settings in your modem's documentation. We used the Intel 144EX's suggested settings and were able to connect at 14.4 and 19.2K to PSI for a relatively good link. We were able to get a relatively higher speed with the Deskporte, at 57.6K for our InterRamp PSI connection. The same

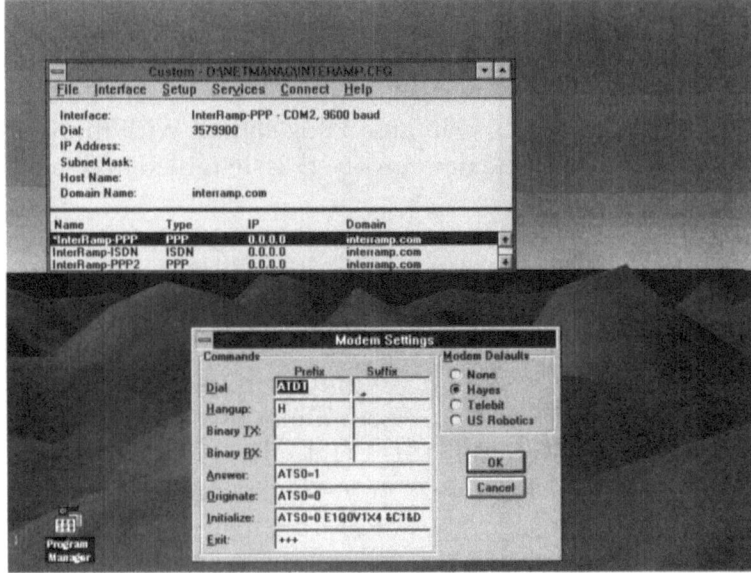

FIGURE 4.3.
Netmanage
Chameleon's
Custom modem
configuration panel.

modems also worked well with Internet in a Box's Dialer application, and they were much easier to configure.

Analog vs. ISDN

The standard phone line system is all you really need to run Mosaic. What you'll find, however, is that it has built-in limitations that will affect the kind of performance you'll get. The standard POTS (Plain Old Telephone System) is still derived from analog technology, and geared for voice transmission. This means that the computer's digital information has to be converted to analog in order to transfer down the POTS line. That's what a standard modem does: it modulates digital computer signals into analog to send them out and demodulates them back into digital as they come in. An analog signal can be considered as an erratic wave formation. The information it transmits is passing over older circuits that were designed in the 1940s for voice, and the translated data has to make the analog journey back and forth in this inefficient format, not to mention the overhead involved in the translation process.

ISDN (Integrated Services Digital Network), on the other hand, uses a digital format from end to end. This means you don't need a modem; ISDN uses what's called a DSU (Data Service Unit) to move digital computer data over digital lines from end to end. This is more like a square wave, with no erratic acoustic patterning found in an analog signal. With more precise digital control over the data involved, a sustained high-speed connection can be made.

ISDN is becoming more and more available in residential versions, and requires little extra equipment. Examples of ISDN equipment that will work with Windows at this time include the ISDN*Tek CyberSpace ISDN card, the QuickAccess Remote from AccessWorks Communications, and the Motorola HMTA 200 hybrid modem (a combination ISDN adapter and V32. modem).

The $395 ISDN*Tek CyberSpace Freedom card is a full-fledged ISDN adapter. It fits into a standard PC ISA slot and has a wide range of IRQ address settings you can select from a set of jumpers on the board. This allows you to avoid conflicts with other cards in your system in the same way we described for the COM ports. You will also avoid having to install and configure COM ports with this board, since it passes information directly to your Windows socket program (in this case, it uses Netmanage Chameleon's Custom application). A Windows-based setup program walks you through configuring and testing the board.

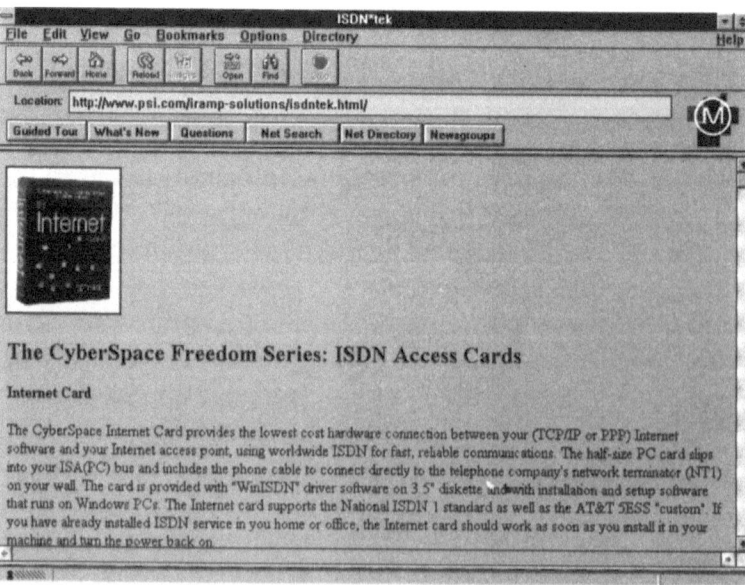

FIGURE 4.4.
ISDN*Tek's CyberSpace Freedom ISDN card page at PSI-InterRamp's Web site.

The only other equipment you'll need with this board is an NT1 adapter. This is a Network Termination unit; it connects your ISDN equipment to the ISDN equipment at your Internet access provider's site. It sits between your PC and your ISDN phone jack. The Northern Telecom NT1 Meridian, roughly the size of a standard external modem and desk or wall-mountable, is offered by Bell Atlantic at $147 and works closely with the ISDN*Tek card, making it a good choice for your Windows ISDN setup. It's becoming more common for ISDN adapters to include Network Termination units built into the base unit, and we expect to see ISDN*Tek produce a card with this capability in the near future. The total hardware cost for the ISDN*Tek card and the NT1 adapter is approximately $550, about twice what you could spend for a fast modem, but affording you true digital connectivity.

ISDN*Tek, PSI, and Netmanage have developed a joint specification for ISDN under Windows known as WinISDN. This specification works around Windows' specific data transfer limitations and allows ISDN to be integrated into the existing Windows socket interface. The CyberSpace card is preconfigured for the Netmanage software and the PSI Internet account service using WinISDN. This allows you to very easily add ISDN to your PSI account.

First, however, you have to have your ISDN line installed. Contact your local phone service to find out if ISDN is available in your area. Use the ISDN profiles provided by your Internet access provider to tell your phone company how to set up your line. In this case we used the profiles provided by PSI to order our ISDN line. The line can be installed either to a new phone jack or in the place of an existing line.

Costs for our example line (San Francisco area, Pacific Bell) ran to $75 for the installation labor, $35 to switch the line from a standard analog line to a digital line, and $15 for the first monthly fee. For $125, we had a digital ISDN line that more closely approximated the high-speed networks that Mosaic was designed to be used with. A caveat to remember is that an ISDN line is not a standard phone line; if you change an old line over, you can't use standard modems, phones, or FAX equipment with it any longer (you can even possibly damage a standard phone by connecting it to an ISDN jack). Special ISDN telecommunications equipment is available, but it's certainly not as common as analog equipment.

Once you have the ISDN line in, you can run ISDN*Tek's Windows ISDN test program to check the line. The next step is to configure your Internet access software, in this case Netmanage's Custom application. Choosing the ISDN option in Custom's startup window changes the pull-down menus slightly; you can now select ISDN hardware under the Setup/Hardware menu. The options are currently tight for Netmanage and ISDN hardware; you can only select the ISDN*Tek card for default configuration. At this point, the configuration goes very smoothly. All you need to enter are the IRQ and memory address settings (from ISDN*Tek's Setup program) and the SPID (Service Profile ID) number provided by the phone company. You're now ready to set up a synchronous PPP link to the Internet. Start your connection in the usual way and you'll be connected to your service provider in ten seconds or less, without the standard modem connection screeches. You can begin to see the benefits immediately: as you start Mosaic, the flow of data will be much faster and home pages will load at a smooth, rapid rate. You can now attempt to access larger multimedia sound, picture, and movie files with more reasonable downloading times.

Another type of ISDN adapter is one that connects to a serial port, like a modem. This would need an open COM port, and would also require a high-speed 16550 UART for stable communications. PSI certifies the $495 AccessWorks Communications QuickAccess Remote as a personal ISDN adapter. It includes an NT1 adapter, so you don't have to pay for additional hardware. It only runs on ISDN lines, and is somewhat limited in its connectivity to asynchronous PPP (analog modem style). While faster data communications are achieved with this unit over standard modems, it uses the same type of data link as a standard modem, so it's not as fast as the synchronous PPP available with the ISDN*Tek card. Its primary advantages are its relatively low cost and the fact that you can move it between different systems and sites easily (for example, between Windows, Macintosh, and Unix computers, and between home and office ISDN accounts).

You can also consider an external ISDN unit that includes a

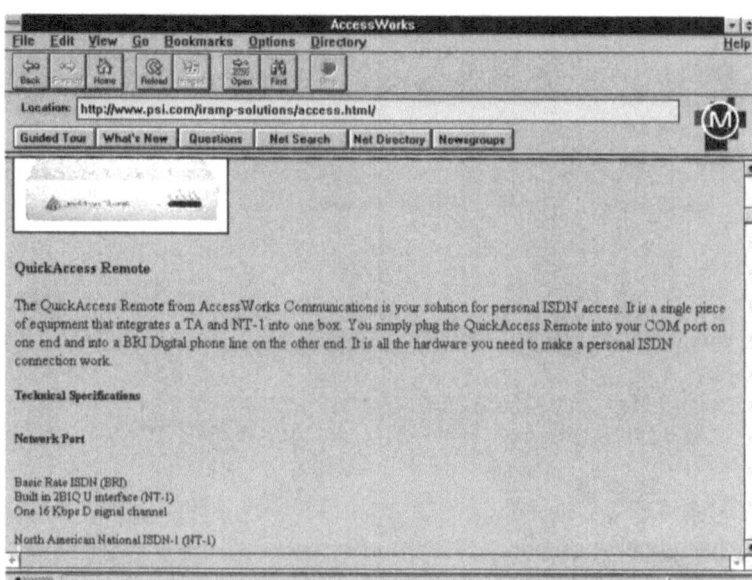

FIGURE 4.5.
QuickAccess
Remote page at
PSI-InterRamp's
Web site.

modem, like Motorola's HTMA 200. This unit combines an ISDN terminal adapter with an integrated NT1 unit and a fast V32 14.4/28.8K modem. It not only works as an ISDN adapter (using your PC's serial port, and subject to the same current limitations to asynchronous PPP as other external ISDN adapters), it also functions as an analog modem over your ISDN line. That means that you can connect to your Internet account via ISDN, and to an on-line service via the modem function. The HTMA 200 also includes a Windows-based Configuration Manager that downloads setup information to the unit.

The hybrid modem is a good solution to the problem of canceling an existing analog line by making it into a digital ISDN line. This unit gives you back the modem functionality thereby lost. Unfortunately, its asynchronous PPP ISDN connection is not the same as a full-on synchronous PPP link. The cost differential (approximately $895 for this unit) would make choosing an ISDN card and an external modem just as reasonable, but you'll have to have two separate lines for that setup to work.

Motorola (and other companies, like IBM) are working on developing a serial port standard for true ISDN under Windows, and we hope to see this available soon. With full ISDN capabilities, external ISDN adapters and hybrid modems would have a better place in Windows setups. For now, however, a card like

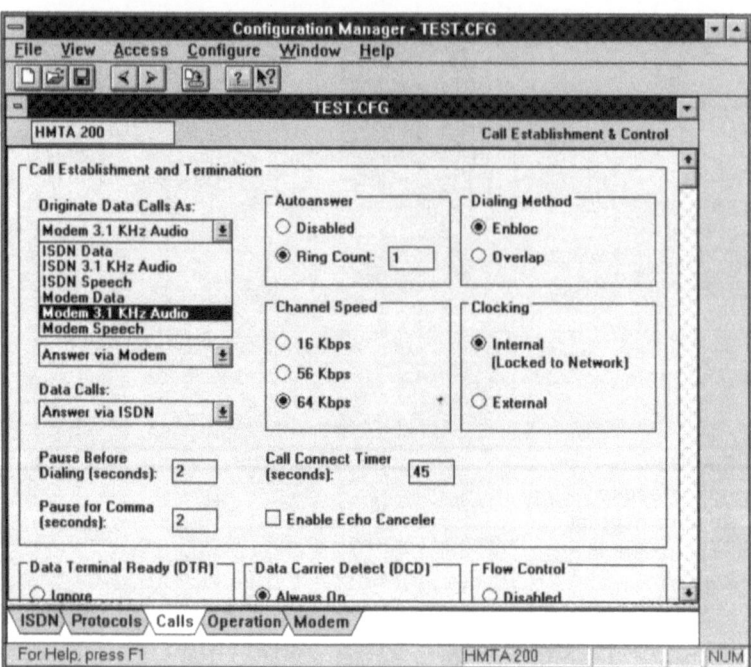

FIGURE 4.6.
Motorola's HMTA
200 Windows
Configuration
Manager.

ISDN*Tek's is your best bet for ISDN connectivity.

It's especially interesting to note how the ISDN service provider, your Internet access provider, the ISDN hardware, and your Internet software all interrelate. In this case, the ISDN service provider brought the line to the PC, configuring the line to match the ISDN adapter card. The card was certified by PSI as compatible with their Internet service, and was listed as being compatible with Netmanage's Internet software. This meant that Chameleon had a driver for the board already included, so configuration was easy. It's important to have this same chain working for whatever Windows ISDN Internet equipment you choose.

5
Venturing onto the Net

You're now ready to venture onto the Internet with Mosaic. The first concept to master is that of the home page/home document.

The home page is a Mosaic document that an Internet site presents as its first link. It's often similar to the table of contents of a magazine. It's also the first document that Mosaic will launch when it starts (the page that Mosaic "homes in on"), and you can set this home page to be any Internet site you can reach. While it's not common to make an FTP directory your home page, you can do it.

The various types of Mosaic for Windows have different home pages and different ways of setting a home document.

NCSA Mosaic connects to a high-use server at the University of Illinois, and presents a home page for the publicly available Mosaic versions (Figure 5.1). This site includes information on the versions of Mosaic available for different platforms as well as Internet search links.

Directly related to the NCSA home page is the associated link to the Mosaic for Windows home page (Figure 5.2). Clicking on that link takes you to a page maintained by the developers of Windows Mosaic at NCSA. The Windows Mosaic home page includes up-to-date technical information, links to the programs and associated viewers (with links to FTP sites), and tutorials.

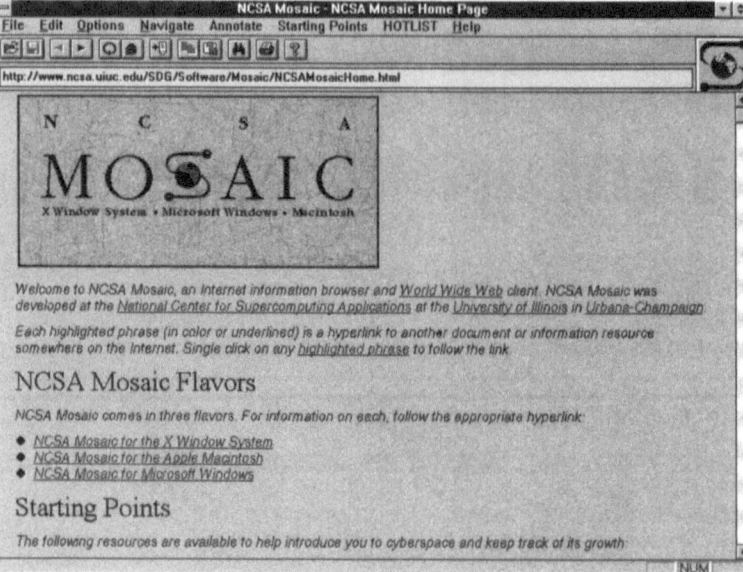

FIGURE 5.1.
NCSA Mosaic
default home page.

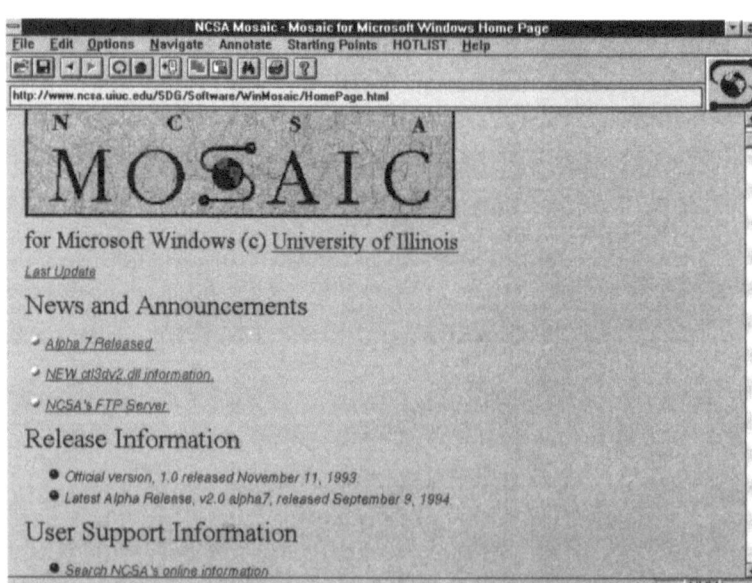

FIGURE 5.2.
NCSA Mosaic for
Windows home
page.

It's a good idea to be able to set your default home document regardless of which Mosaic or other Web browser version you're using. Remember, the NCSA home page loads over the network each time you start NCSA Mosaic. This means that you're starting a network transfer from the University of Illinois to your computer at that time. There's no reason to impact that site so heavily when you can easily set your home page to another place, and you might not want to wait for the NCSA document to load, either, when you really want to travel somewhere else.

To set the home page in NCSA Mosaic for Windows, either change the relevant section of the General Preference panel under the Options menu (for Alpha 9 and above), or edit the Mosaic control file (for earlier versions). Do the latter by starting the Notepad accessory and opening the Mosaic.ini file.

Right at the top, under the [Main] section, are the lines:

```
Autoload Home Page=yes
Home Page=http://www.ncsa.uiuc.edu/SDG/Software/Mosaic
/NCSAMosaicHome.html
```

The first line is the switch that will prevent NCSA Mosaic from loading a home page automatically. Change this to

```
Autoload Home Page=no
```

to start Mosaic by itself. You'll still be directly connected to the Internet (as long as your Windows stack is still running), and you'll be able to navigate using the menus and the hotlist, but you won't connect to Illinois every time you launch the program.

Rewrite the second line to load an alternate page as your home document. We'll set this to the Mosaic for Windows home page as an example. Change it to

```
Home Page=http://www.ncsa.uiuc.edu/SDG/Software/WinMosaic/HomePage.html
```

Save the file, restart Mosaic, and it should now link directly to the Mosaic for Windows home page.

You can also use Set Mosaic, the Unofficial Mosaic Configuration Sheet (Figure 5.3), to set NCSA Mosaic's home page. Just enter the page you'd like to link to on the home page line, and Set Mosaic will write this information to your Mosaic.ini for you.

It's best to use these techniques to move your home page to the least traveled site possible; the Windows Mosaic example should

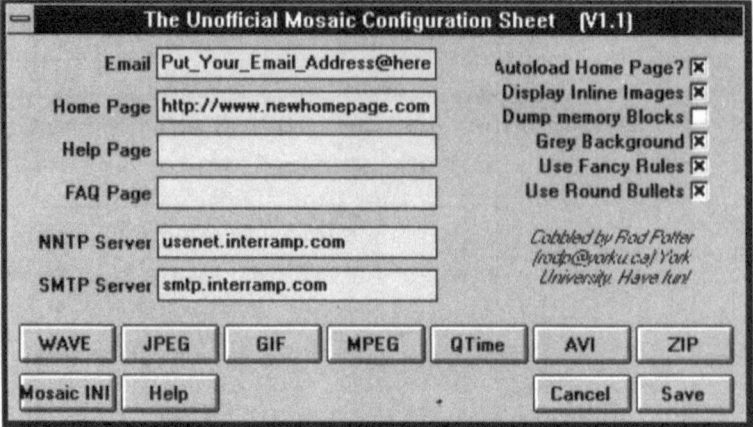

FIGURE 5.3.
Set Mosaic
Unofficial Mosaic
Configuration sheet.

be used sparingly, and that goes for the default NCSA page as well. Once you're comfortable navigating menus and hotlists, turn off the autoload home page feature, or link to your own favorite place directly.

Spyglass's Enhanced NCSA Mosaic loads an initial document at start-up that is local. This document has links to the Spyglass home page on the Internet, as well as to several files on your hard drive that are set up during its installation.

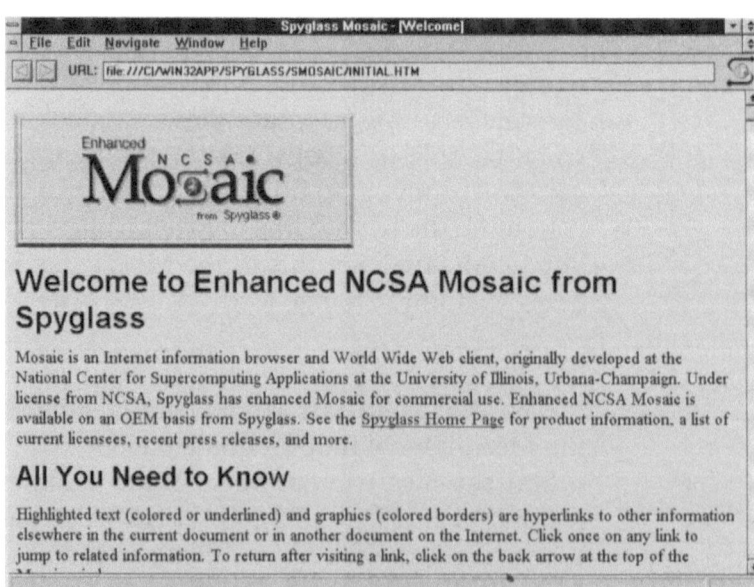

FIGURE 5.4.
Spyglass Mosaic
home page.

This means that the home document loads very quickly and reliably, as it doesn't have to come over the network. Spyglass Mosaic has a help system built into a Mosaic hypertext document that is local as well, and using this is a good way to get comfortable with hypermedia and hypertext links while off-line.

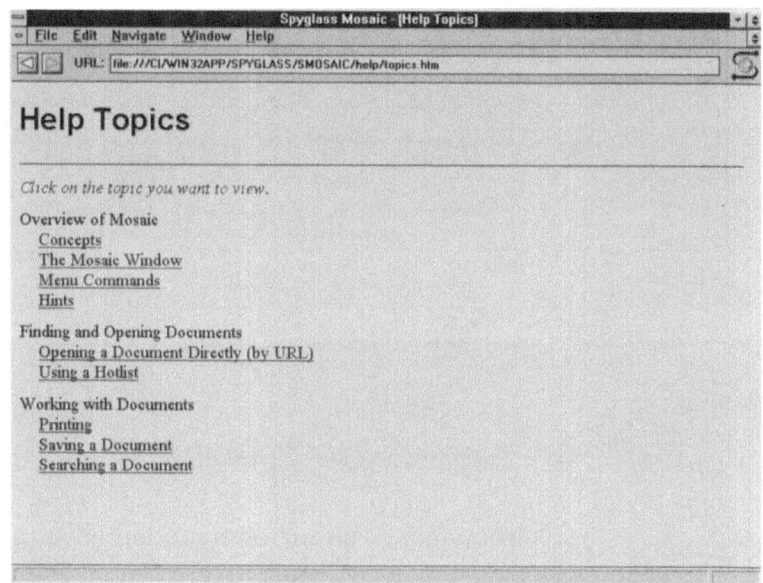

FIGURE 5.5.
Spyglass Mosaic's
in-line hypertext
help system.

To change the default home page, use the Preferences panel under the Edit menu, and change the text in the box labeled Home Page to the site you prefer.

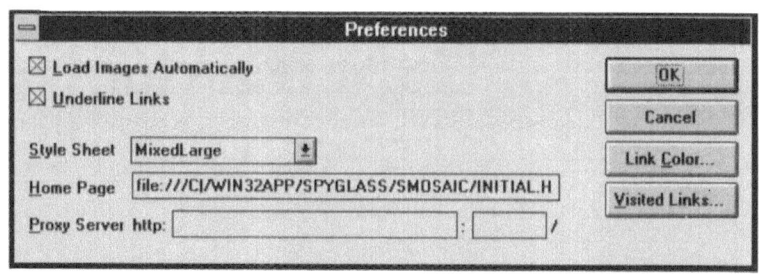

FIGURE 5.6.
Spyglass Mosaic's
Preferences panel.

The Initial page also has CyberSpace sampler of hypermedia links to the Internet, stored locally, as well as a fast link to the Spyglass home page. The Spyglass Web site features up-to-date product information and press releases about Enhanced Mosaic, and is a good place to find out about updates and new releases of the software.

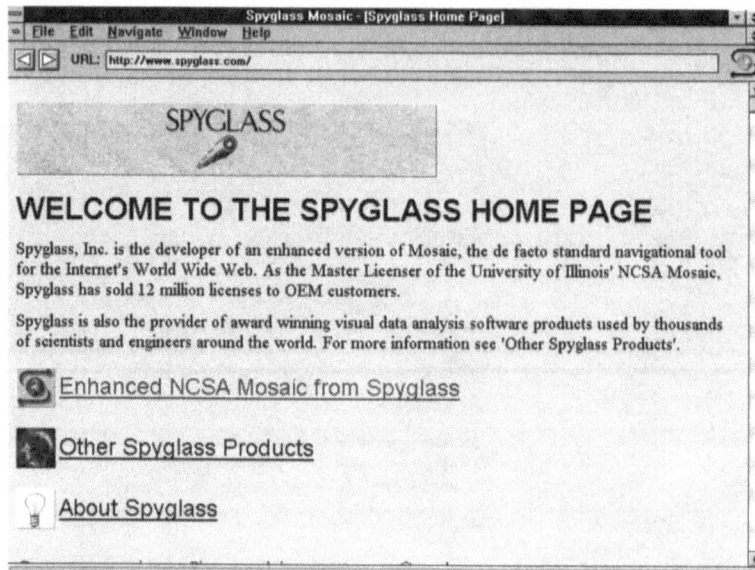

FIGURE 5.7.
The Spyglass
Internet home page.

Netmanage's WebSurfer, included with their Internet Chameleon package, is a Mosaic-style browser that also includes a local home page.

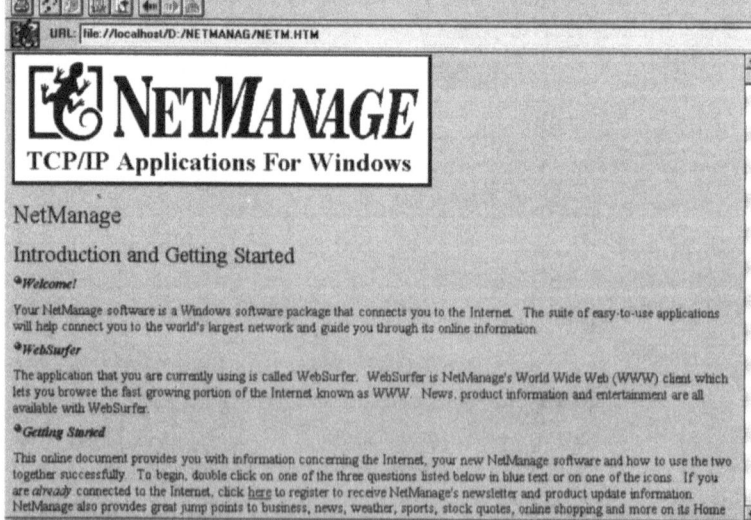

FIGURE 5.8.
Netmanage
WebSurfer's default
home page.

Set WebSurfer's default home page (called a "Start-up Document" in this release) by using the Preferences panel under the Settings menu.

Netmanage's linked Internet Web page is a good source for information on their products, including the Internet connectivity tools mentioned in this book.

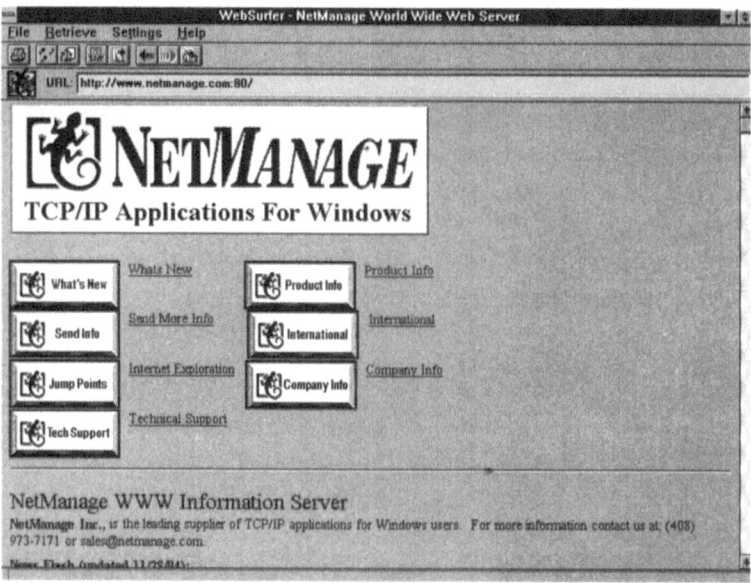

FIGURE 5.9.
The Netmanage Internet home page.

Internet in a Box takes a full-scale approach to the home document concept. IBox's AIR Mosaic starts up by loading the GNN (Global Network Navigator) home page. This is a highly interconnected Internet site with links to several places of interest presented in a unique digest format.

GNN is a very good way to get introduced to the Internet. It's exceptionally well connected and informative. You can also use it as a home page in any other version of Mosaic (for example, if you're tired of NCSA or Netscape Communication's Web site); just enter the line http://gnn.com/gnn/GNNhome.html as the address for your new home page.

Setting the Home Page in AIR Mosaic is easy. Under the Options menu, the Configuration panel has the same kind of Home Page entry as in Spyglass Mosaic, and it also includes an option box for turning on and off the automatic loading of this page at start-up.

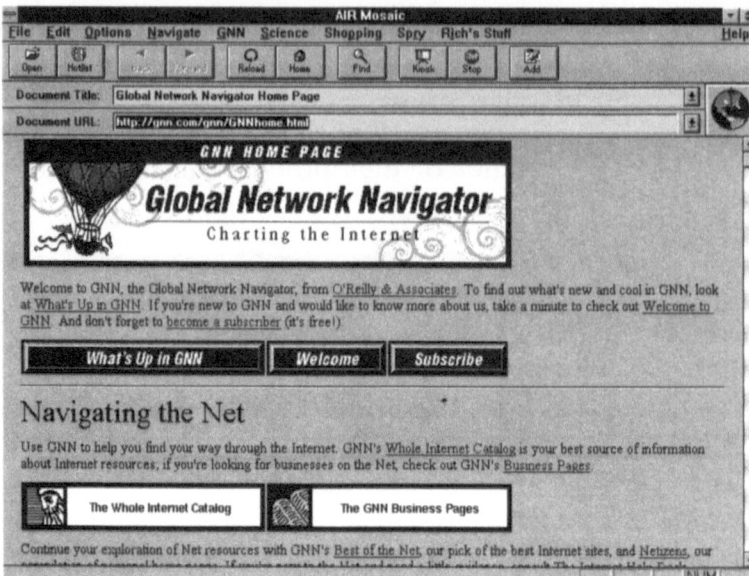

FIGURE 5.10.
AIR Mosaic,
launching GNN as its
home page.

Netscape Communications Corporation's Netscape has a default home document that links to their colorful, informative home page. This is a good site for Netscape users, as it includes information specific to this version. It also features Internet links and an integrated help system. This differs from the Spyglass version, where the help system was stored locally; in Netscape, accessing the help menus from inside the program will link to the main Mosaic Communications server without even having to go through the home page. You will have to have your Internet connection up and running to get this help, however.

Setting an alternate Home Page in Netscape is a function of the Preferences panel under the Options menu. Click on the drop box icon to the right of the top listing and scroll to the Styles item. This will access the Styles page, which includes a Window Style box. Edit the "Start With:" line with the address of the site you'd like to load automatically each time you start Netscape. Click on either the Blank Page or the Home Page Location button to turn the automatic link on and off.

Starting WinWeb lands you at the EINet Galaxy, a clearinghouse for Internet information that's arranged a bit like GNN. This is a good place to start looking into the Internet.

Among Windows Web browsers, WinWeb has the easiest way to set a home page. Under the Options menu, the Set Home Page

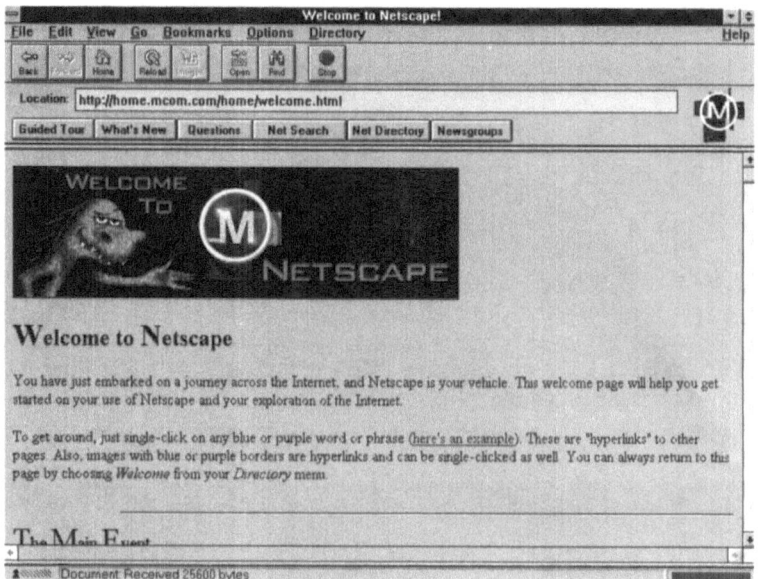

FIGURE 5.11.
Netscape loading
the Netscape
Communications
Internet home page.

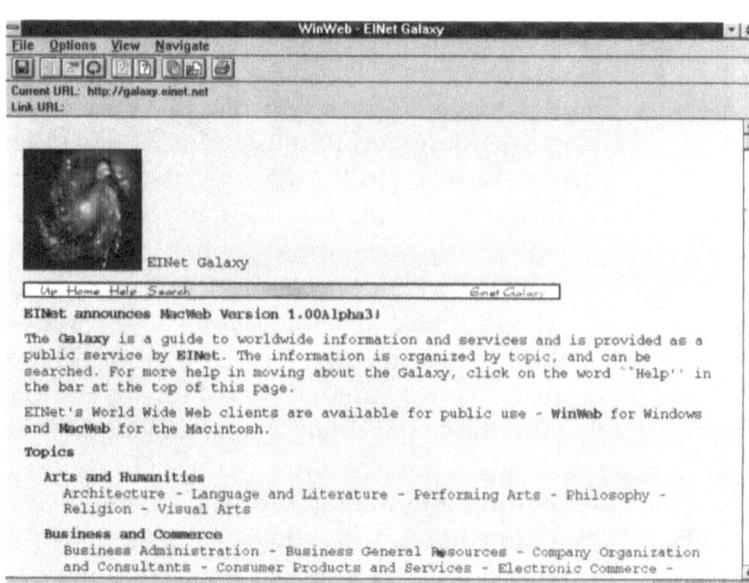

FIGURE 5.12.
WinWeb at the ElNet
Galaxy home page.

item will launch a small panel that you can use to switch the default home page from the EINet Galaxy to the current page you're at, or enter in a different listing.

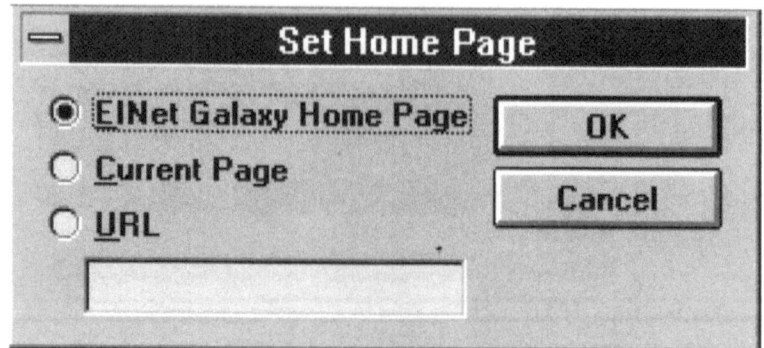

FIGURE 5.13.
WinWeb's Set Home
Page panel.

URL Protocols

Mosaic uses a technique called TCP/IP to transfer information over the Internet. This information comes in a variety of formats, including text, pictures, compressed binary computer programs, and sound files. It's stored in many different ways as well, including file archives (FTP), gopher indexes, and Mosaic's own HTTP (hypertext transfer protocol). The URL (Uniform Request Locator) was designed to allow Mosaic to act as a central clearing-house for these different formats. That means that Mosaic can read hypertext documents in its own HTML (hypertext mark-up language) format, access gopher indexes and FTP sites, and cvcn read Network Newsgroups all within its main program window.

The first URL protocol type you'll come across is HTTP. Your default home page is in this format. This protocol allows the transfer of documents with links to remote Internet sites and picture and sound file links embedded in them. When you click on a highlighted picture icon, HTTP negotiates between your computer and the remote server computer, requests the picture, and downloads it to your local system. HTTP also handles links to other Internet sites via hot-linked words or pictures. HTTP's main format is HTML. HTML is a file type; it's designed to give Mosaic the foundations to build a hypertext document out of text and picture files found on the Internet (or even on your local system, as in the Spyglass help system).

A sample of this can be viewed in NCSA Mosaic for Windows. Bring up the default NCSA Home Page. Next, select the Document Source item under the File menu.

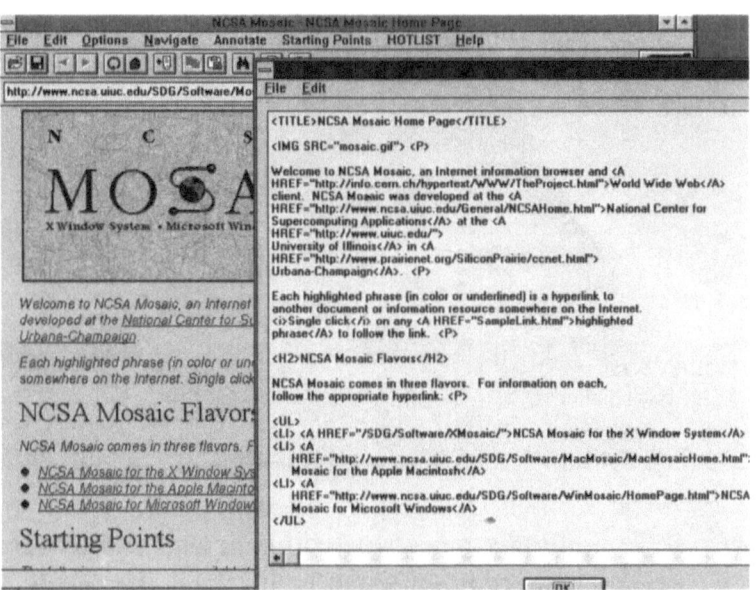

FIGURE 5.14.
NCSA Mosaic and source HTML code.

Note how the page is constructed of the body text with marked pointers to hypertext information and Internet sites. Mosaic directly translates these tags into the related links and images and loads the page in the more-familiar Mosaic format, using HTTP.

Other URL types include FTP, gopher, and NNTP (network news transfer protocol). Mosaic can also load some file types directly; for example, you can load .TXT files directly into any Mosaic version's main window. AIR Mosaic and Netscape 0.93 (and above) also have a drag-and-drop feature that allows you to drag a file (HTML, text, or graphics) into their main windows, where it will load automatically. You can also drag a file directory from the File Manager into these programs, and use it to navigate your own system. Netscape also loads .JPG and .GIF files directly into its main window, while AIR Mosaic will launch an external viewer for these types of files, even when using drag-and-drop.

Most of the URL types that Mosaic can access directly are preconfigured. This means that they'll work with your version with-

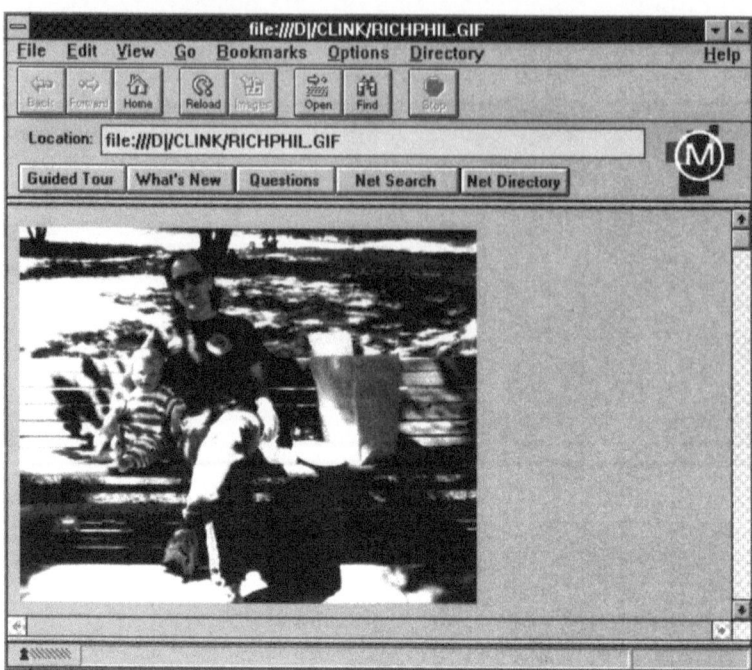

FIGURE 5.15.
Netscape's in-line image capabilities (local GIF file).

out any more work on your part. A link on the NCSA Windows Mosaic home page to the FTP site where the files are located will bring up an FTP archive automatically, for example. One exception is NetNews; Mosaic can only access the worldwide network discussion groups and NetNews file archives directly if you have a valid News server. This is a part of a good Internet access provider account, so make sure your access provider gives you the name of a News server that you can configure properly. At this time, only Netscape Communications' Netscape is preconfigured to run Network News well.

You can use the other versions of Mosaic to read News; it'll just take a bit of configuring. Internet in a Box is already halfway there; SPRY provides a default News feed for use with AIR Mosaic. Type news:* at the Open URL line, and you should get a group listing of SPRY's NetNews feed (Figure 5.16).

Note the format; the numerical listing to the right of the newsgroup is a code corresponding to the number of messages posted at a given time.

SPRY's News server censors out some of the more controversial newsgroups that are available, however. The prior censor-

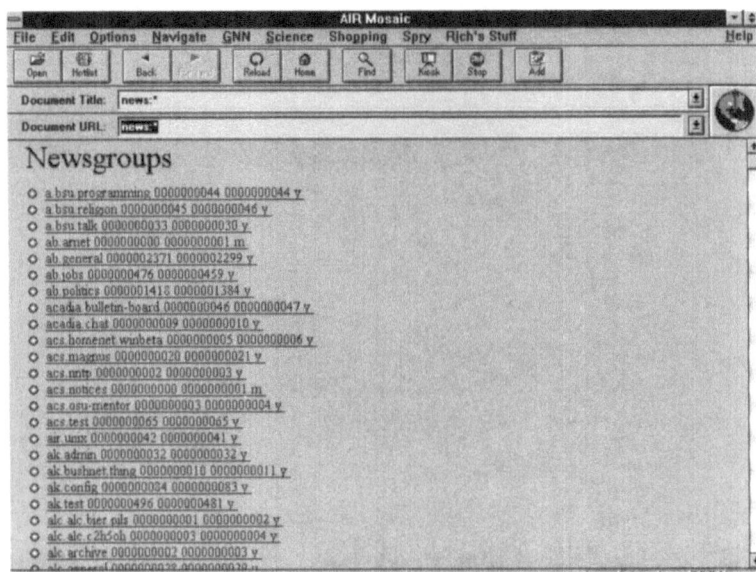

FIGURE 5.16.
AIR Mosaic as a
NetNews reader
under the SPRY
news feed.

ship of the more controversial newsgroups is disheartening. We can change the default News server from the default **SPRY** News feed to a more comprehensive feed from **PSI**, our Internet access provider. Make sure your Internet access provider gives you a full News feed.

PSI has provided us with a News server address as part of our password-protected Internet access account. Accessing our carefully preserved documentation, we see that our NNTP News server at **PSI** is called `usenet.interramp.com`. We can now open up AIR's Configuration panel (under the Options menu), and put this line into the News server field (replacing `news.spry.com`). We should now be able to access the much richer News feed from **PSI** directly in AIR Mosaic. Use Open URL from the File menu, and type in `news:news.announce.conferences`. The resultant newsgroup will look like Figure 5.17.

Now we have a full news feed, but it's still only really useful for browsing newsgroups. Unfortunately, you can't post easily with most versions of Mosaic.

You can also load newsgroups to the public domain versions of Mosaic from NCSA by editing the relevant section of the `Mosaic.ini` file.

In our example, we've changed the line under [Services] to reflect our NNTP server. Note that the News feed listed here isn't

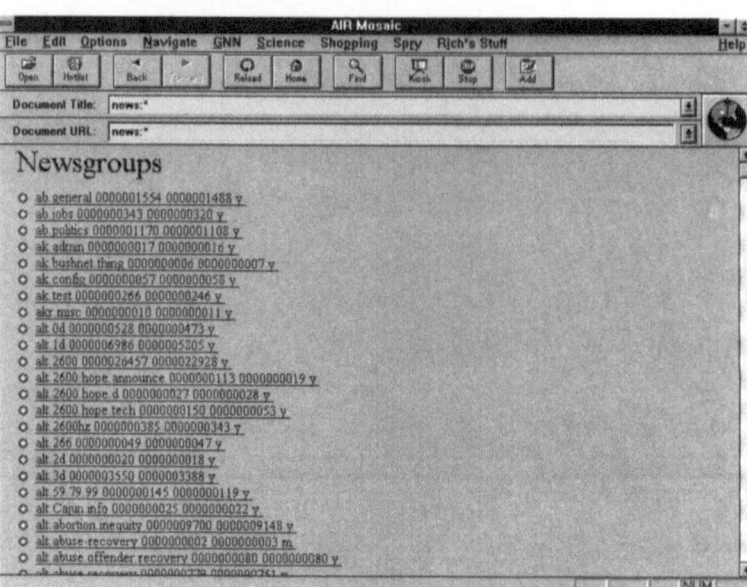

FIGURE 5.17.
AIR Mosaic as a
NetNews reader
under the PSI news
feed.

public access; it won't work unless you have a PSI InterRamp
account.

```
[Services]
NNTP_Server=''usenet.interramp.com"
```

Replace that line with the name of your NNTP server, which
should be available from your Internet access provider.

Save the file, and you can now reach newsgroups under NCSA
Mosaic. Selecting Open URL from the File menu and entering
News:* gives us a similar listing as with AIR Mosaic.

You can use this method as a fairly reliable browser with other
versions of Mosaic (like Spyglass Enhanced NCSA and Netmanage's WebSurfer) by editing the relevant NNTP server entry in
their configuration screens, but it doesn't take the place of a full-
featured News reader.

Netscape Communications, on the other hand, has taken steps
to provide a better News reader interface for Mosaic, one that
rivals external News readers and still works from inside the Web
browser. You get a full News feed, the ability to move between
threads (links between postings), and an interface for posting and
responding to messages directly from the Netscape interface.

You can only use this News feed with Netscape, because it's
provided from Netscape Communications themselves, and not

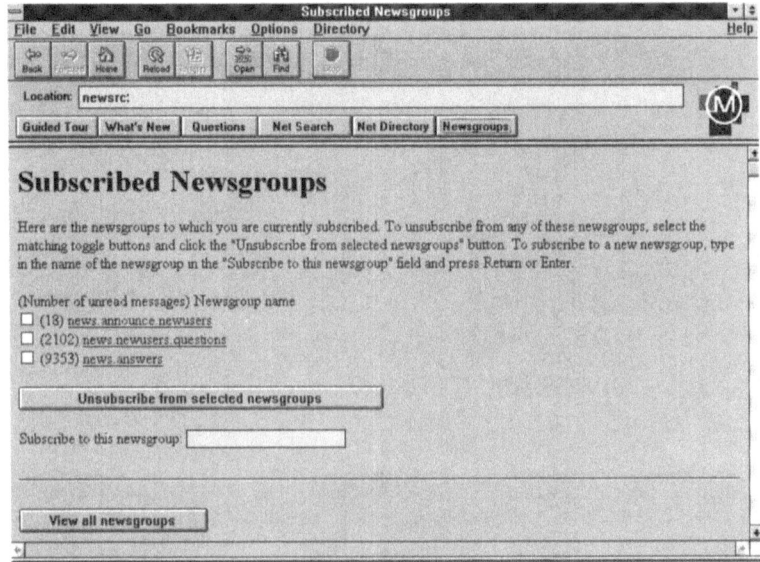

FIGURE 5.18.
Netscape's initial
News screen.

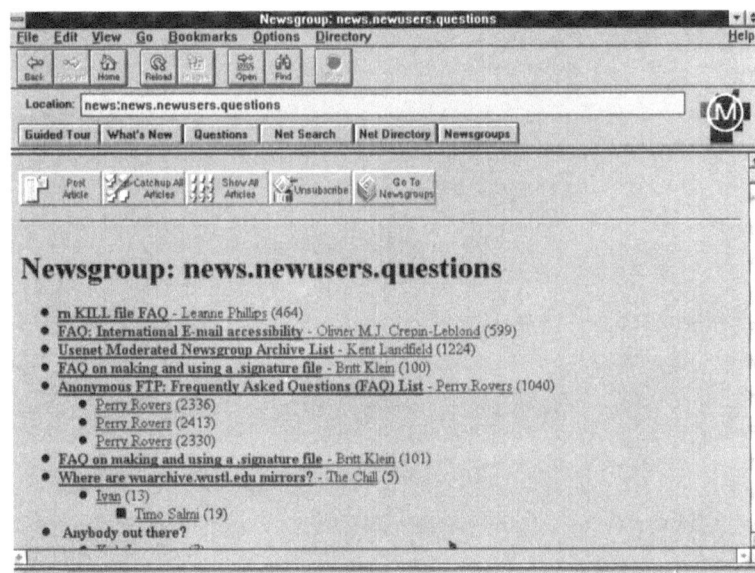

FIGURE 5.19.
Netscape's News
reader, showing a
list of Network News
postings for the
`news.newusers.`
`questions` group.

your Internet provider. This could be a good way to get access to Network News even if your access provider doesn't give you a News server. Just use Netscape to get there (but note that the Release 1.0 version of Netscape no longer features the Newsgroups button, as does the earlier versions, and you'll have to use the directory menu item, or enter the URL by hand).

6
Gopher: The Worldwide Internet Encyclopedia

Gopher is a way to search for information over the Internet. The gopher protocols were developed to help manage the sheer volume of information present.

Gopher servers were designed for use long before Mosaic-style Web browsers were developed. A base-level gopher is still accessible from a terminal shell type of Internet account.

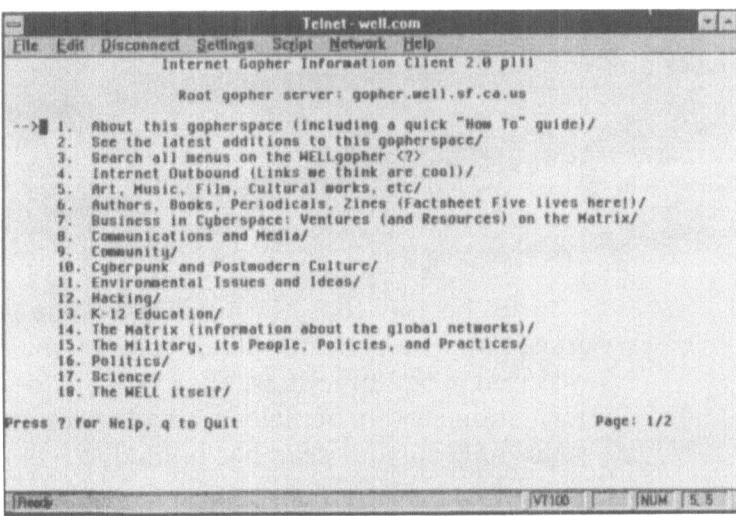

FIGURE 6.1.
The text-based gopher at the WELL in Sausalito.

Once the World Wide Web was developed, however, new types of gopher searches had to be developed, to take advantage of the nature of the Web and its powerful browsers. To this end, search engines like Archie, Veronica, and Jughead became standard ways to access information using a variety of intelligent software query types.

What this means to the general Mosaic for Windows user is that they can access information on the Internet by looking it up. These interfaces are all available for use with the versions of Mosaic and Netscape that we've covered in earlier chapters.

NCSA Mosaic provides a number of points to access gopher servers right from its main screen. The Starting Points menu has a pull-down submenu of gopher servers at various locations. If we go to the Gopherspace Overview at the top of the submenu, we land at a server at the University of Michigan that gives a good index of the various gopher servers available for us to browse through.

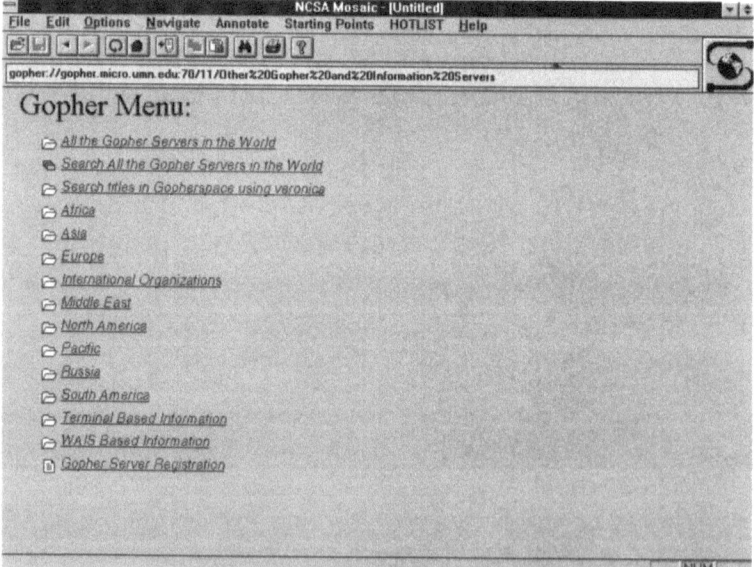

FIGURE 6.2.
NCSA Mosaic main window, showing the Gopherspace Overview at the University of Michigan.

Note the way that Mosaic displays the gopher information; it's set up as a series of file directories that will list gophers in different geographic sites. The rationale for this is that these information servers actually are in the countries listed. It makes sense to keep your searches limited to a close geographic area, since you'll be passing commands along to the computer you want to search, and sending those requests internationally can be inefficient and wasteful.

We can go directly to the subdirectory for North American gophers, and then to the listings for the United States, for our example. Note that we're now looking at a series of directories for servers in various states. If we wanted information specific to an area, we could go directly to a gopher near there.

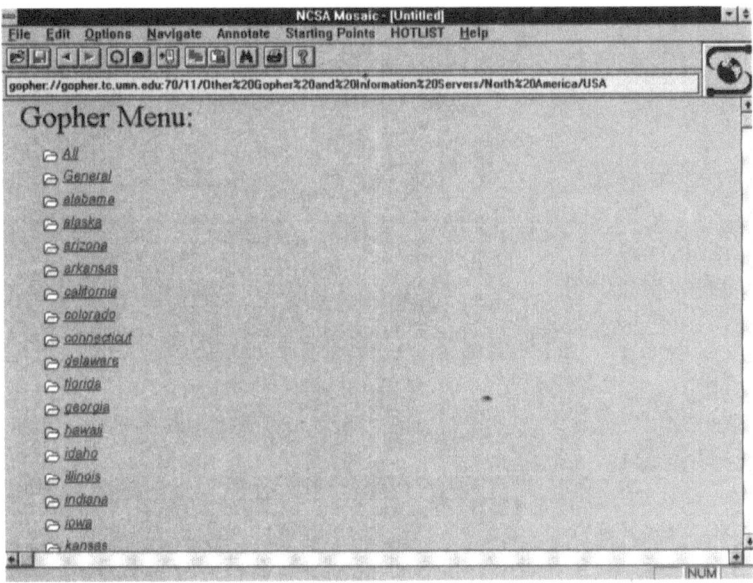

FIGURE 6.3.
United States gopher directories.

For example, to find out information specific to California, we'd select that list item, and we'd be able to browse gophers in California (Figure 6.4).

Note how the information is laid out. Each line is a link to a gopher search engine at a different site. Mosaic's hypermedia interface is used to get to each of the gophers listed just by clicking on the relevant link.

We can get to the gopher for the California Department of Education (Figure 6.5) by following the link here. It shows us information in a variety of different formats, laid out by the staff who implemented the server.

Besides subdirectories containing specific information, there's a search engine on this page. If we go to the Search All CDE Menus link, we'll get a screen where we can look up information

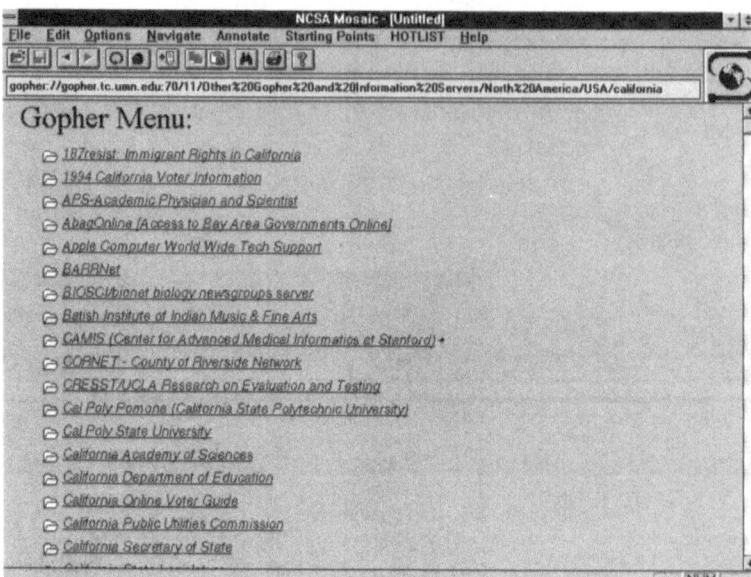

FIGURE 6.4.
Gopher servers in
California.

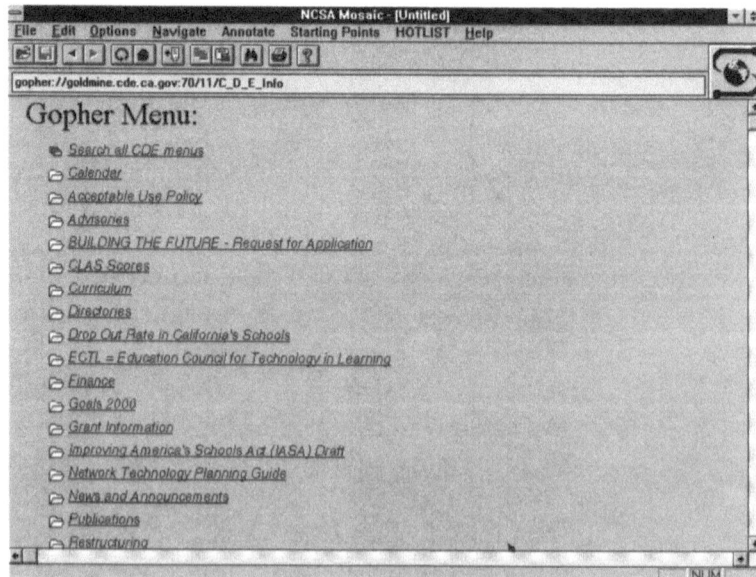

FIGURE 6.5.
The California
Department of
Education gopher.

on this server. We tried a sample search for Alameda County, and got back 1993 results on California Learning Assessment System scores for the county in a large spreadsheet.

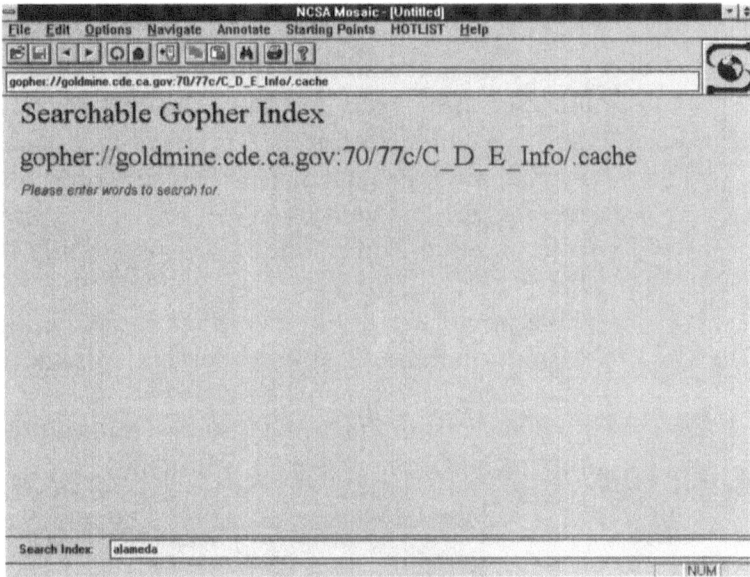

FIGURE 6.6.
The California Department of Education's gopher search screen.

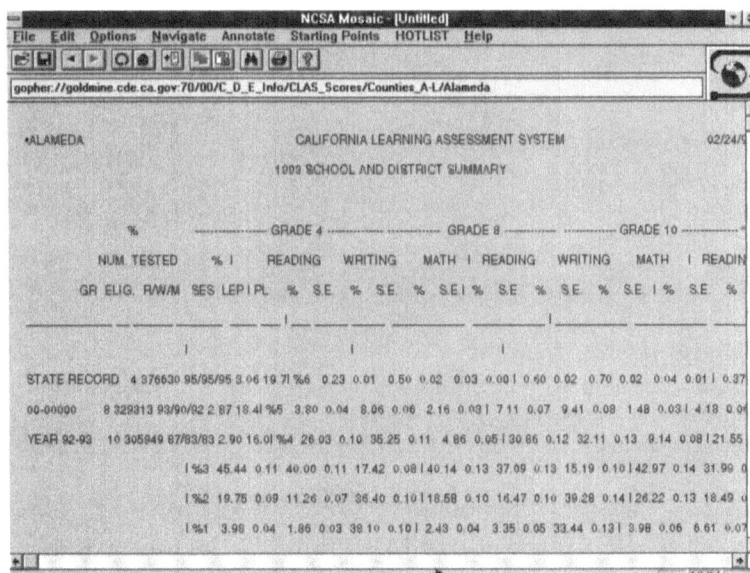

FIGURE 6.7.
Results for the search on "alameda" at the California Department of Education's gopher.

This is just a small sample of what you can do with a basic gopher; with a little bit of judicious digging about, you can get very close to the information you're looking for. Gophers are key to the World Wide Web because they tie in the large range of text-based retrieval systems with the Mosaic hypermedia interface.

You should also note that a gopher is a form of URL, and just like the other type of URLs that Mosaic can access, you can get to a gopher directly. Try accessing Microsoft Corporation's at `gopher://gopher.microsoft.com` by typing the line into the Open URL or Open Location dialog box, and you'll bypass any other menus. You can also save any gopher listing to your hotlist file.

Veronica works like a gopher, but it can also access more information than just gopher servers, including HTTP Mosaic Web pages and FTP directories. There's a Veronica search point accessible under the Starting Points menu under NCSA Mosaic, at the top of the Gopher Search submenu.

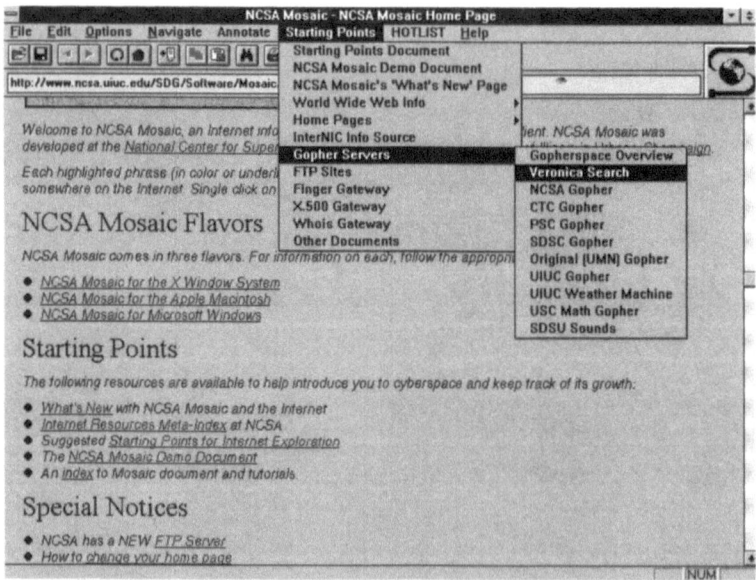

FIGURE 6.8.
NCSA Mosaic's route to the Veronica search engines.

This leads to a series of Veronica search engines laid out geographically, like the gopher example given above; try to access the one closest to your site. The interactive Veronica engines listed at the top will do this for you, and will also help to make sure you make a connection. We used the link that would select a server for us, and would search through all gopher data types.

Our test search was on "joystick," a common piece of PC equipment. We found several items listed, including description files, beta test announcements, equipment for sale, calibration programs, and more. These listings are all links to information on computers at other sites, linked together by the Veronica search engine.

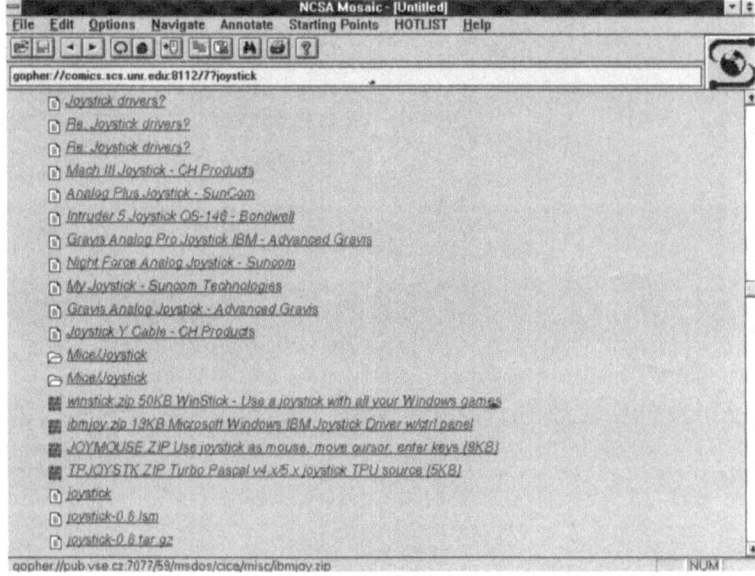

FIGURE 6.9. Veronica search output, showing text files, directories, and programs.

Note how the icons for the different data types reflect their source type. The text files have a text icon, the binary program files have a binary icon, and the directories are shown as file folders. We can read the text files directly in Mosaic, and download the programs by setting the program to Load to Disk (or use shift-click in Netscape).

Web Searches

Various standards bodies and independent organizations have also developed some interesting alternatives to traditional gophers that make more full use of the Mosaic interface. The forms capability of the latest Mosaic-style Web browsers (versions over

2.0 for NCSA) mean that you can access multiple search engines at a single site.

The CUI (Centre Universitaire d'Informatique) W3 Catalog search engine is just such a site. Available at the link from the NCSA Mosaic home page called Internet Resources Meta Index (at the top of the next list), it's a global entry point into World Wide Web information. You can also get to it directly at `http://cui_www.unige.ch/w3catalog`.

It's very easy to use: just type an entry into the text field, and hit the submit button. We looked up the word *planets* for our example.

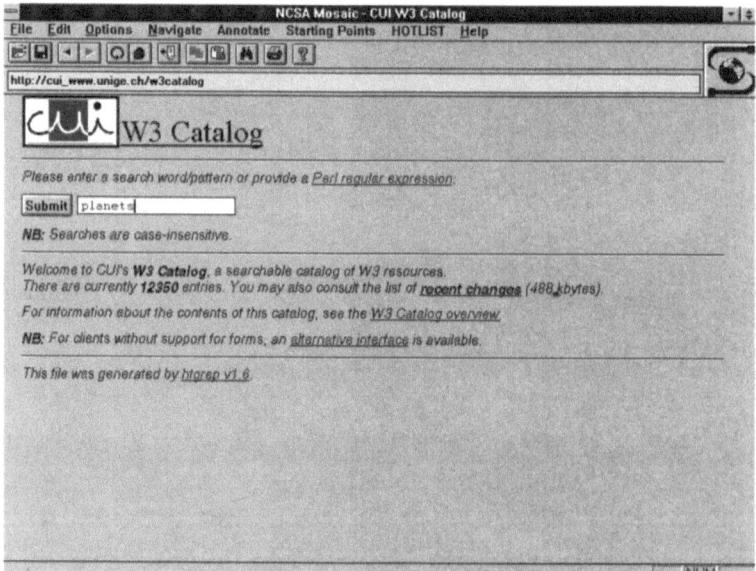

FIGURE 6.10.
The CUI W3 main search page.

This search engine not only looks up information, but displays its result in a digest format, with hyperlinks built into the results page. Our search on *planets* returned several sites with planetary information, ranging from the Planetary Society to NASA laboratories. It also located a fan-organized United Federation of Planets from TV's "Star Trek" (located at `http://www.imall.com/startrek/ufp.html`)!

It's up to you to determine just how useful the items linked in a search results page like this will be for your needs. Try exploring particular links to see the wealth of information available. We went to the first link, at the Planetary Society, and found a society

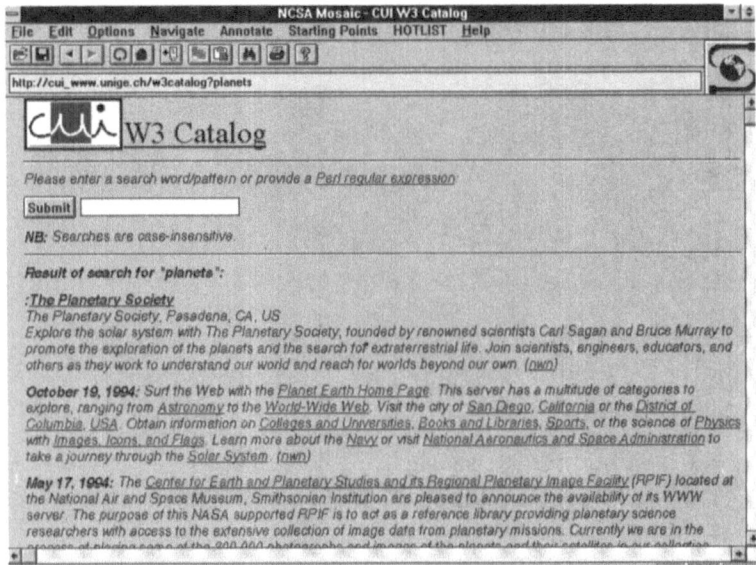

FIGURE 6.11.
The results on the
World Wide Web
information search
on the word *planets*.

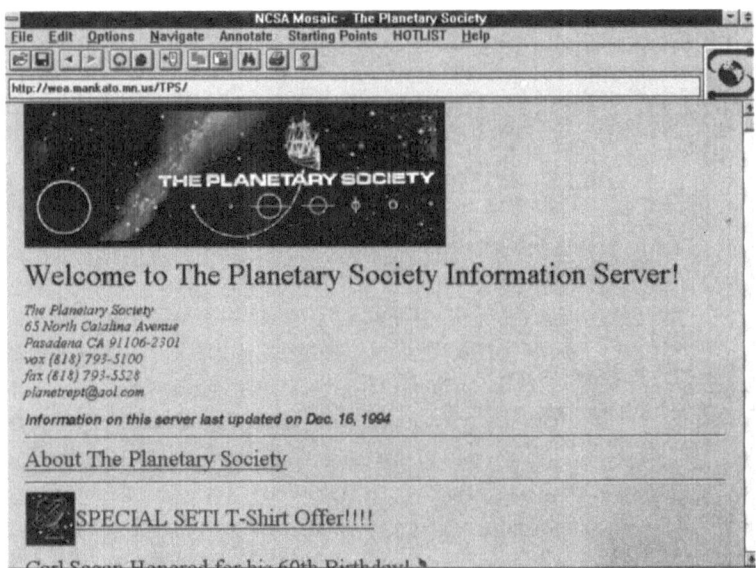

FIGURE 6.12.
The Planetary
Society home page.

enthusiastic about the exploration of the solar system.

Their Internet site has links to large stores of planetary data, including multimedia tours of the solar system. We followed one of these links to a site in Arizona, and really found what we were looking for.

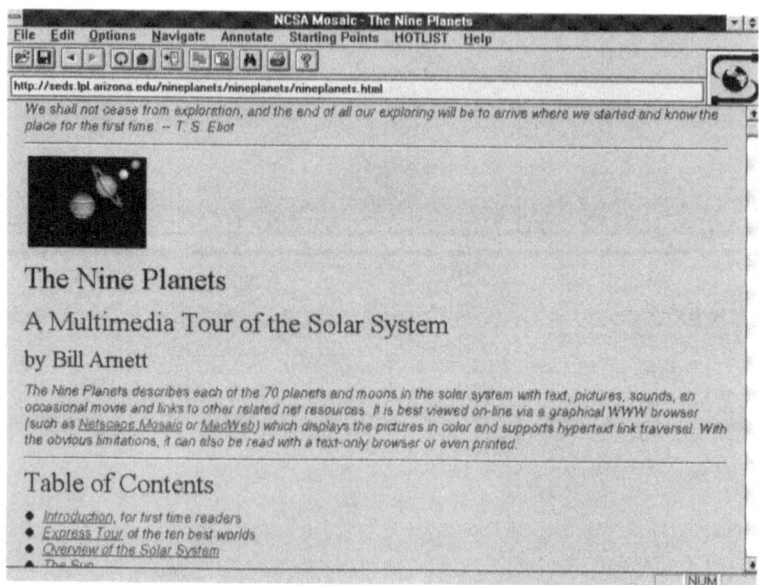

FIGURE 6.13.
The Multimedia Tour
of the Planets.

Bill Arnett's Multimedia Tour of the Planets (`http://seds.lpl. arizona.edu/nineplanets/nineplanets.html`) is a comprehensive, well laid out overview of the solar system through complex maps and images, sound files, and text documents. We continued our search, now focusing on the planet Mars, and selected its name from the Multimedia Tour list.

We got a comprehensive list of Mars information, including an opening page with a beautiful picture of the planet, a list of facts, and links to more pictures at other sites. Note that pages with this much information can be quite long, and will take some time to download to your computer.

This is a great example of how the Internet can be used as an on-line encyclopedia, with its own search engines and hypermedia links. Adding a Planets section to your hotlist, and adding the results of a search like this, means you'll only have to do the W3 search once, and you'll have the links to use directly from your Web browser.

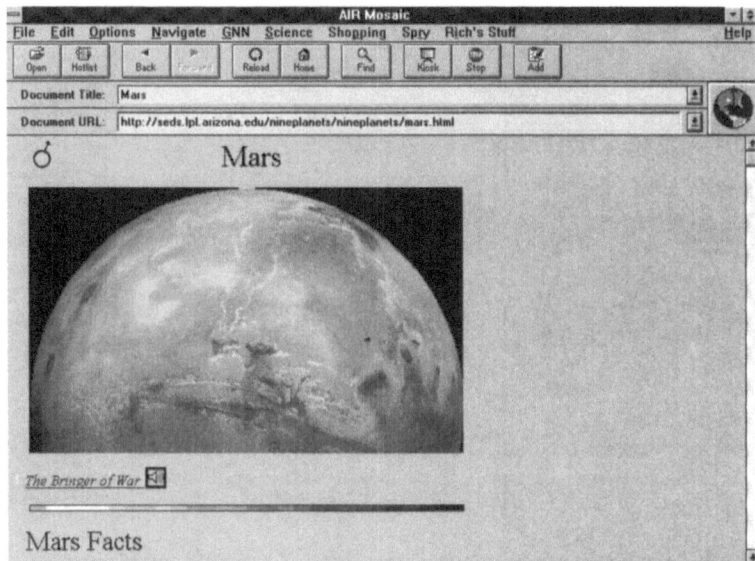

FIGURE 6.14.
The Multimedia Tour of the Planets' Mars page, top, with opening graphic and embedded sound file link.

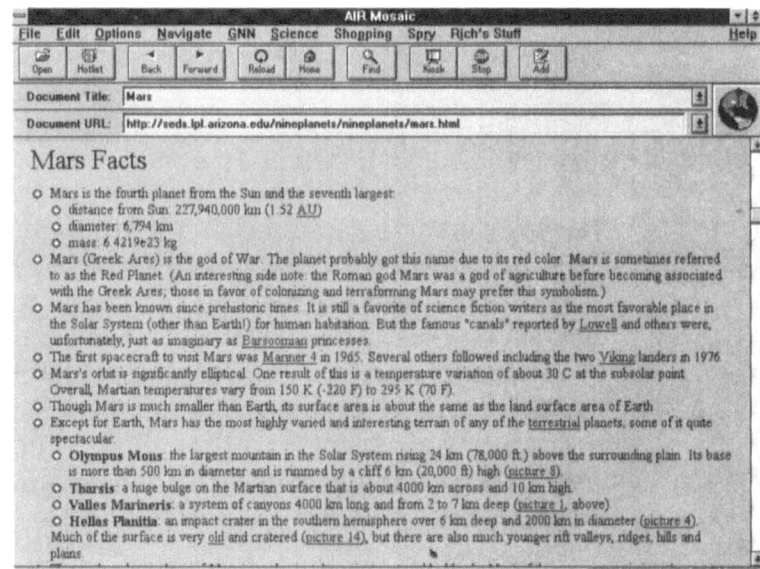

FIGURE 6.15.
The Multimedia Tour of the Planets' Mars page, middle, with hyperlinked facts section.

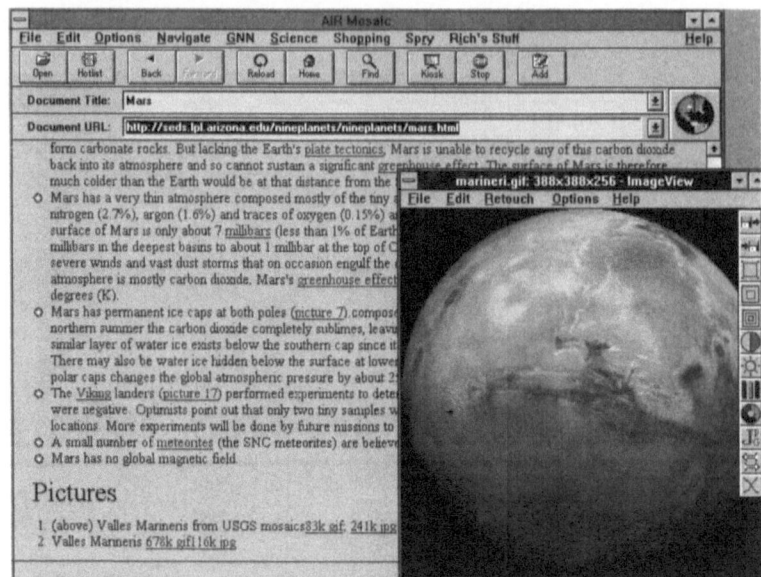

FIGURE 6.16.
The Multimedia Tour of the Planets' Mars page, lower half, with picture files, links, and an external viewer launched.

There's also an experimental meta-index at NCSA that you can use in Mosaic; this provides a way to search more than one type of index at the same time from one point. Note how many of our previously mentioned search types are listed on this one form, including a Veronica search and the W3 search page. This page can be accessed at http://www.ncsa.uiuc.edu/SDG/Software/Mosaic /Demo/metaindex.html (Figure 6.17).

Advanced Searches with Netscape

Netscape takes the approach of the NCSA Meta Index a step further, and provides an easy link to several advanced search types from its Net Search page. You can access this from the button bar just by clicking on Net Search (Figures 6.18 and 6.19).

This brings up a well-indexed page of search tools, with helpful examples and descriptions. Some of these include Web crawlers, specialized search engines that can find information on the Internet via descriptive indices that the programs construct themselves.

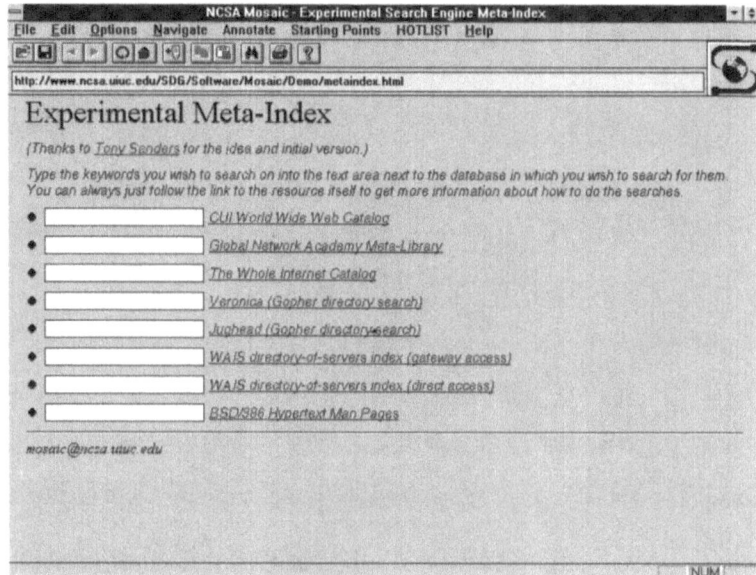

FIGURE 6.17.
Experimental
multipart search
form at NCSA.

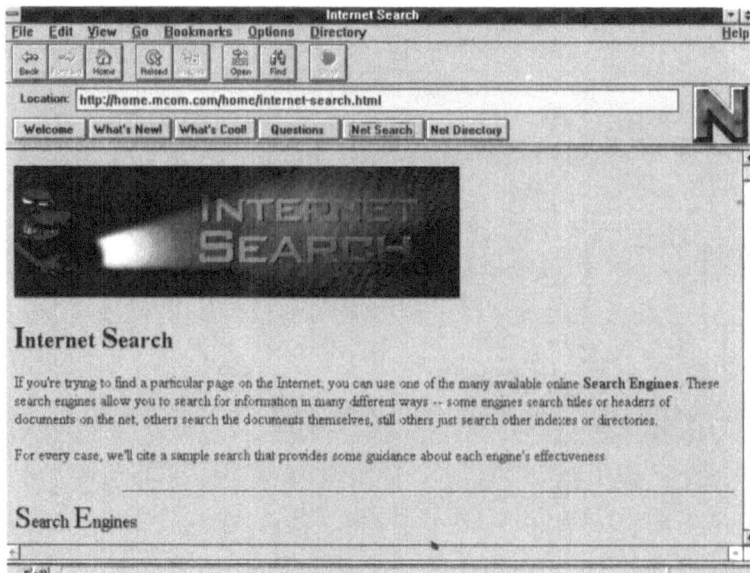

FIGURE 6.18.
Netscape's Internet
Search opening
page.

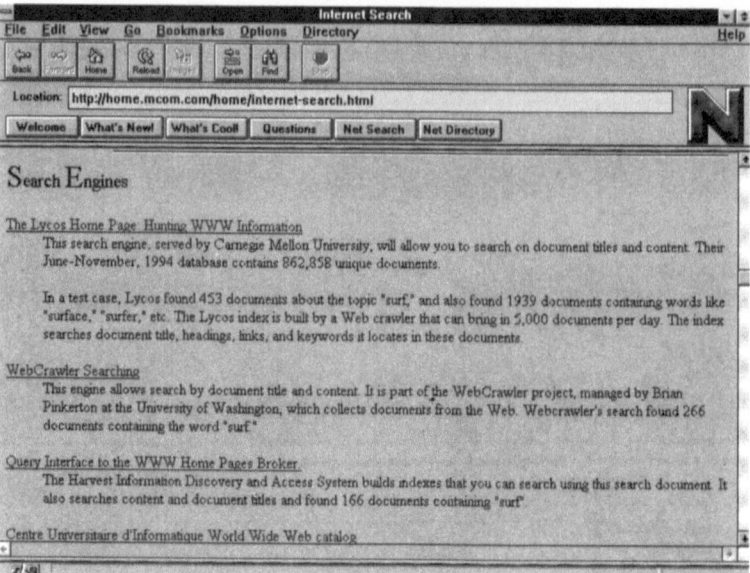

FIGURE 6.19.
Search engines at
Netscape's Internet
Search page.

The LYCOS Web crawler at Carnegie-Mellon University maintains a list of over one million Internet references, and it's growing. It uses software robots to gather information for its database by actively looking over the Internet for new references. We can set the search parameters in a number of different ways, including limiting the number of responses and the extent of the database we want to search (the smaller database search will process faster).

We decided to try a large search of the entire database, and went to one of the search forms. It presented us with a panel similar to the CUI search engine, and we placed our query there.

A search on the word *pregnancy* found 164 items, and gave us a digest of the first 20 results (Figure 6.22). These included Network News items, Web sites, and online medical journal items. The index is hyperlinked, and accessing the information from your search is easy.

FIGURE 6.20.
LYCOS Web Crawler
opening page.

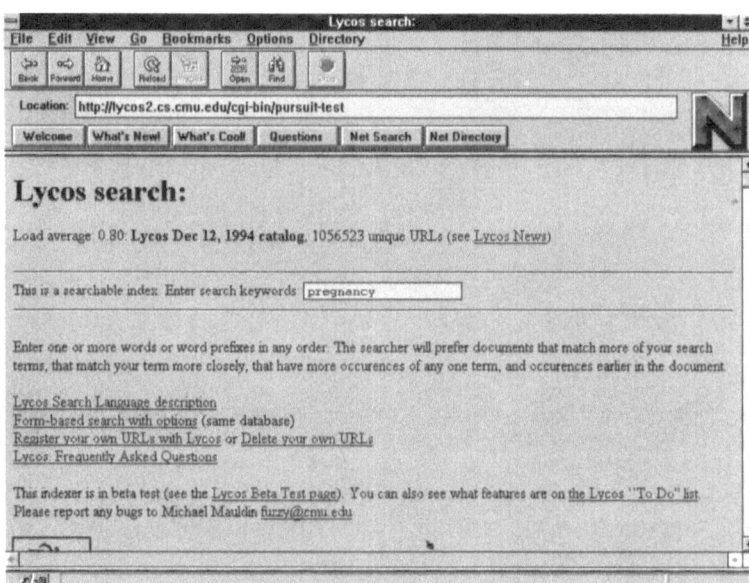

FIGURE 6.21.
LYCOS's search
page.

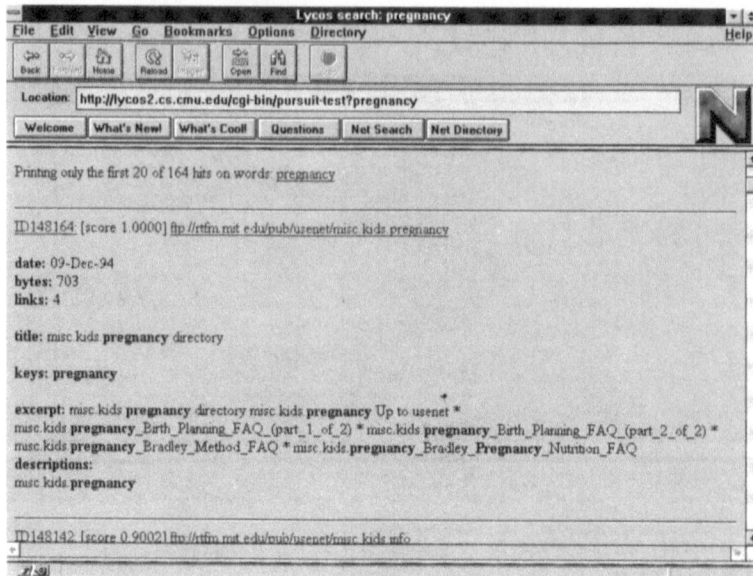

FIGURE 6.22.
LYCOS search
results on the word
pregnancy.

You can use the other Web crawler links on Netscape's Internet Search page in the same manner. It's also possible to make a hotlist of search tool items from both the NCSA sites and those listed in other Web browsers like Netscape.

7
Files on the Net

FTP, the File Transfer Protocol, is a common way to transfer programs, text, picture, and sound files over the Internet. An FTP site is accessed in almost the same manner as you would access your own hard drive. FTP handles transmitting the commands from your system to the remote site, sending a request to move to another directory, for example. Mosaic can use FTP directly, and makes accessing FTP sites as easy as using the Windows File Manager.

For this example, we'll access an FTP site with a large collection of PC shareware files at the University of Washington. Using NCSA Mosaic, pull down the Starting Points menu, and select FTP Sites (Figure 7.1).

This will take you to a large index of FTP servers, listed alphabetically, with descriptions (you can also get there from other Web browsers by going to `http://hoohoo.ncsa.uiuc.edu:80/ftp-interface.html`). We've looked for a good PC shareware site, and found `wuarchive.wustl.edu`. To get there, just click on the blue link (Figure 7.2).

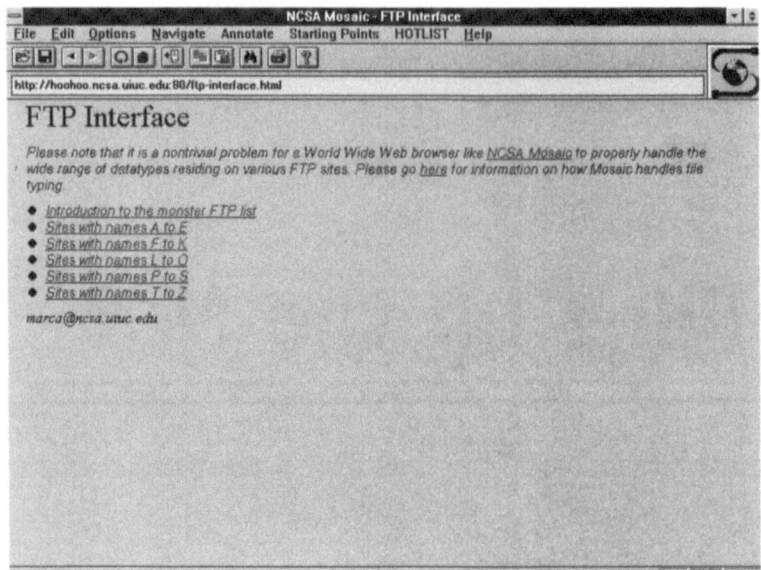

FIGURE 7.1.
The FTP Interface
main screen under
NCSA Mosaic.

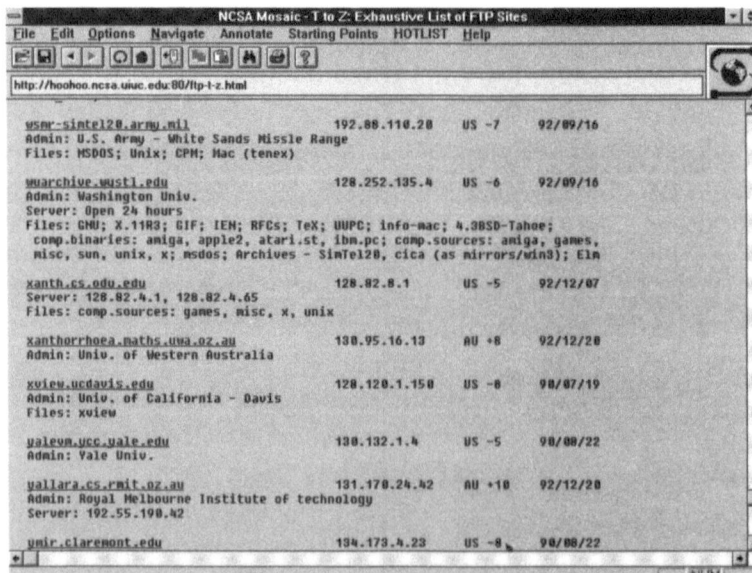

FIGURE 7.2.
A section of the
FTP Sites list under
NCSA Mosaic.

After connecting, Mosaic will show you a hypertext view of the remote file system. Note that the URL address listing has changed; the remote file system isn't a Mosaic-style HTML document but an FTP site. Mosaic can handle the different data types on the fly, so you don't have to worry about this.

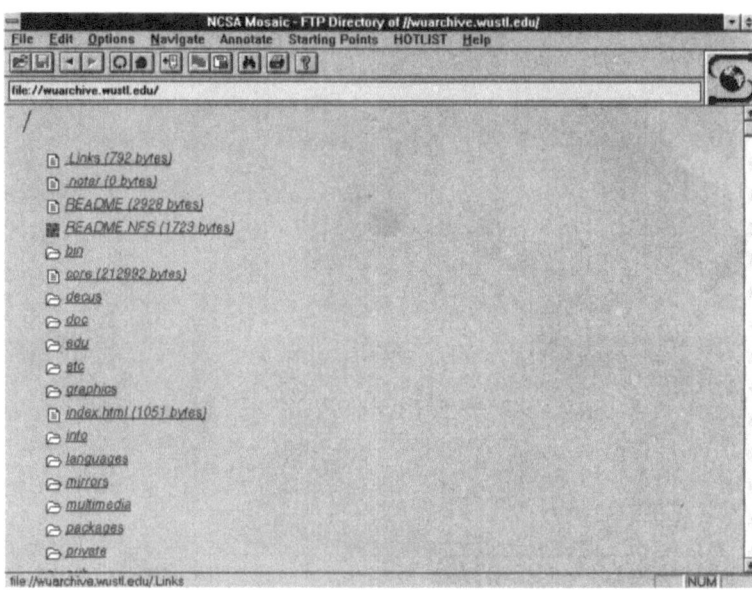

FIGURE 7.3.
FTP archive at
wuarchive.wustl.edu.

You can now move around the remote file system in a manner similar to using Window's File Manager. Each item is linked; directory links will bring up a view of their contents. Moving down into the /pub directory, we start to see program files (Figure 7.4). Mosaic is now differentiating between the different data types (text files and binary programs) it has located by identifying them with different icons. It uses a series of 1's and 0's for the binary programs and a text file icon for text items, just like with gopher views.

We've explored a bit further, and found a Windows 3 directory with potentially interesting files hidden under cryptic names (Figure 7.5). Look for a text file that relates to the binary ZIP files; these are usually paired up. Clicking on the text file will load it directly into Mosaic, where you can read it before you go to the trouble of downloading the file.

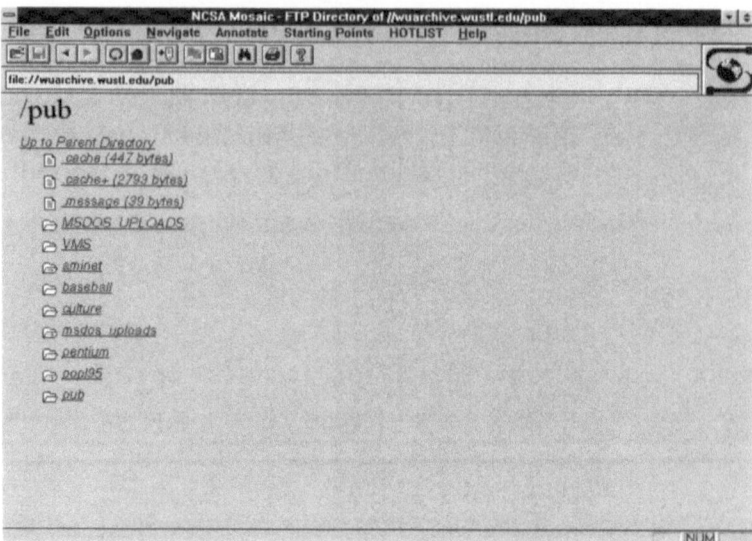

FIGURE 7.4.
The /pub directory at
wuarchive.wustl.edu.

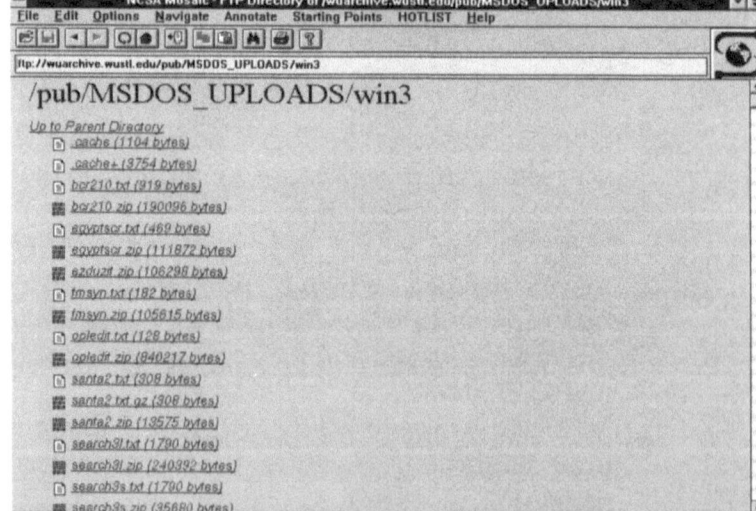

FIGURE 7.5.
The
/pub/msdos_uploads/
win3 directory at
the University of
Washington archive.

For our example, we found the files bcr210.txt and bcr210.zip.
To find out more information, we just clicked on the text file (the
one with the .txt file extension). It then loads into Mosaic:

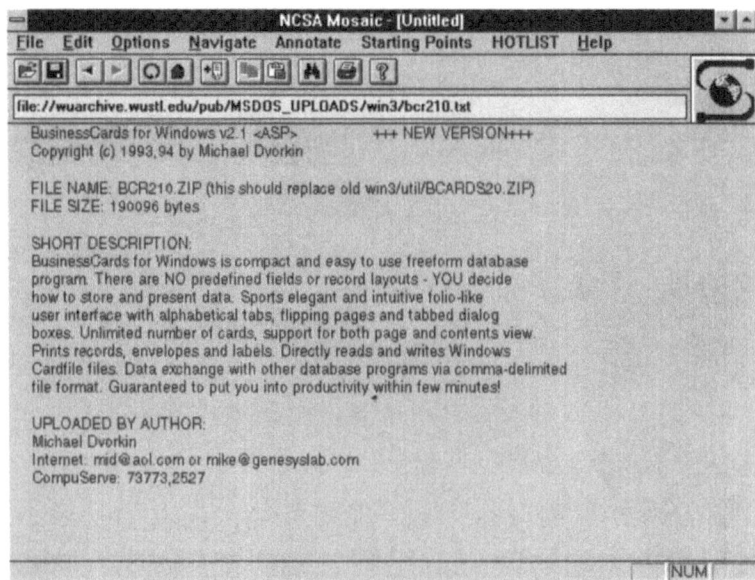

FIGURE 7.6.
An example text file loaded from an FTP archive into Mosaic.

To download the file, just move back one screen, set Mosaic to Load to Disk (under the Options menu), and click on the .ZIP file name. You'll get a dialog box asking you which directory to save the file in, and Mosaic will then transfer the file to your hard drive.

Archie

For a more sophisticated FTP search tool, try Archie. Archie looks for files and file archives, and can find items by name. For our example, we went to the NCSA Mosaic home page (`http://www.ncsa.uiuc.edu`), and selected the Internet Meta Index. Interactive Archie programs are among the subjects listed, and we used those links to go directly to the Archie search pages (Figure 7.7).

The large Archie index at Nexor in the United Kingdom is provided as a public service, and is a good place to start looking for files. We went to the opening page, where we could see an index of Archie search programs listed by country. You can reach the Nexor Archie index directly from other Web browsers by going to `http://web.nexor.co.uk/archie.html` (Figure 7.8).

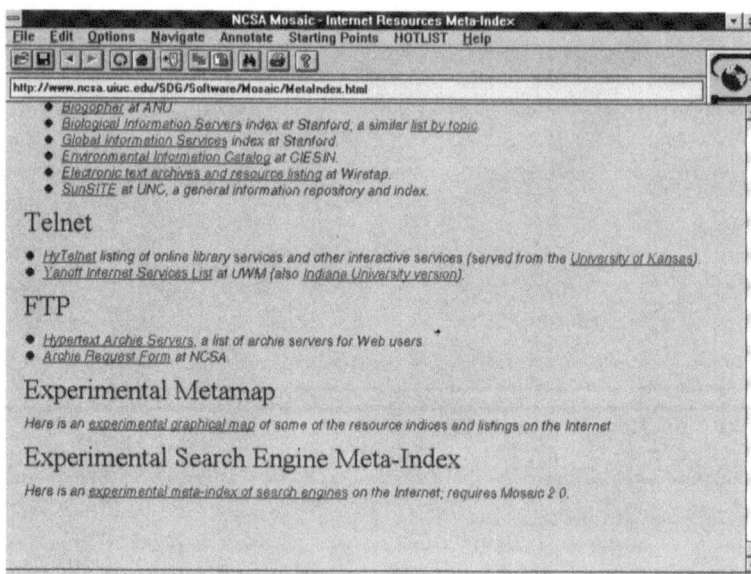

FIGURE 7.7.
The Internet
Resources Meta
Index listing for
Archie servers.

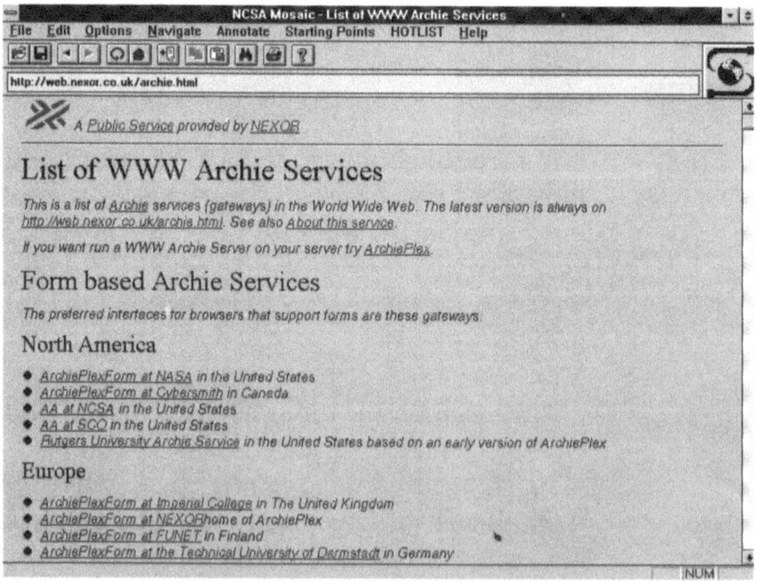

FIGURE 7.8.
The Archie index at
Nexor.

We picked an Archie server in the United States, at NCSA, and went to the search page for it.

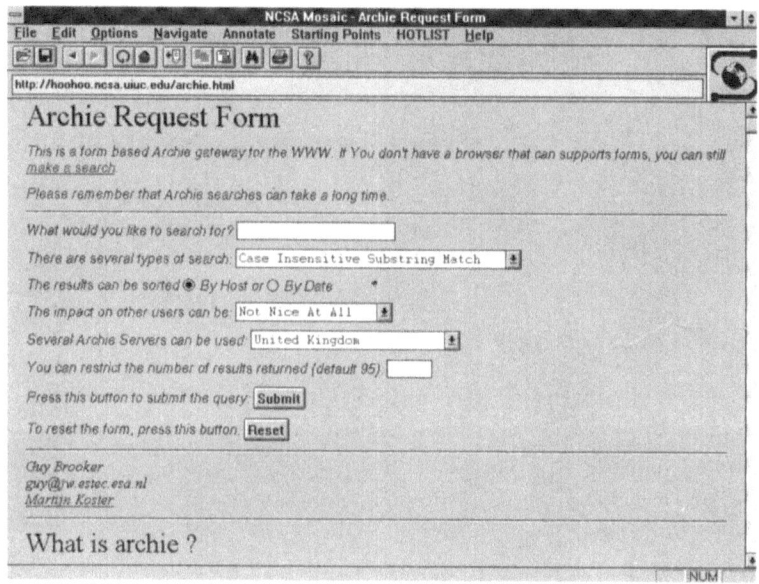

FIGURE 7.9.
The Archie server at NCSA.

At this point, you're presented with an interactive search form along the same lines as the gopher and Veronica search engines discussed earlier. Use the form to limit your search range and type, the number of responses, and the specific Archie server to use (listed geographically). We tried a representative search on *cdrom*, entered it in the top panel, and got a list of directories and files that matched our search string.

Note that the gopher-style icons for the Archie results page don't show up in the Windows version of NCSA Mosaic, and generate an error icon instead. This won't affect your search, and the hypermedia links will still work. Other Web browsers, like Netscape, will show the icons correctly. We switched to Netscape for the remainder of the example (Figure 7.10).

We browsed the result listing to find Windows and DOS related files, then used the links built into the Archie index to move to those files and directories. The listing for a msdos cdrom archive gave us direct access to an FTP directory at a server in the United Kingdom (Figure 7.11).

We can then read the index file directly in our Web browser, for descriptions of the contents of the archive. Notice that the

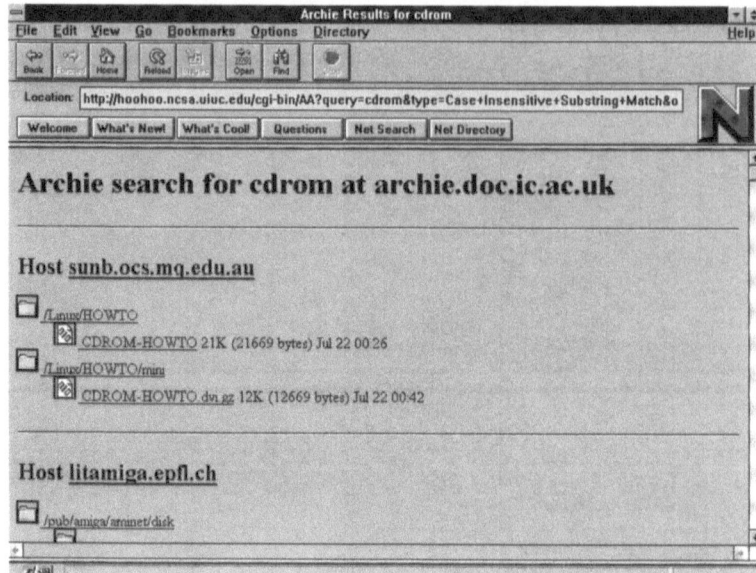

FIGURE 7.10.
Representative
Archie results for
the term *cdrom*.

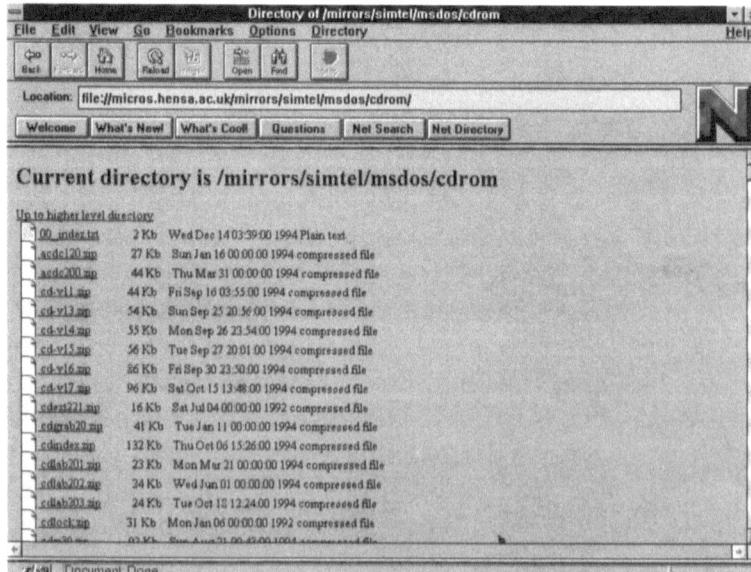

FIGURE 7.11.
A MS-DOS cdrom
archive listing.

index has a text file icon, so we know we can display it directly.

To download the cdrom utilities in Netscape, shift-click on the icons. Set Mosaic and other Web browsers to Load to Disk first.

Galaxy Search

Another way to search for files on the Internet is to use a good interactive index. The EINet Galaxy home page has a search function built right into it. That's the home of the Web browser Win-Web, and you can reach it as the default home page when you start that program.

To get to the EINet Galaxy from another Web browser, go to http://galaxy.einet.net. At the bottom of the home page, you'll find a search panel. We entered the term *shareware*, and restricted our search to the EINet Galaxy itself (you can also do a World Wide Web and Gopher search from here).

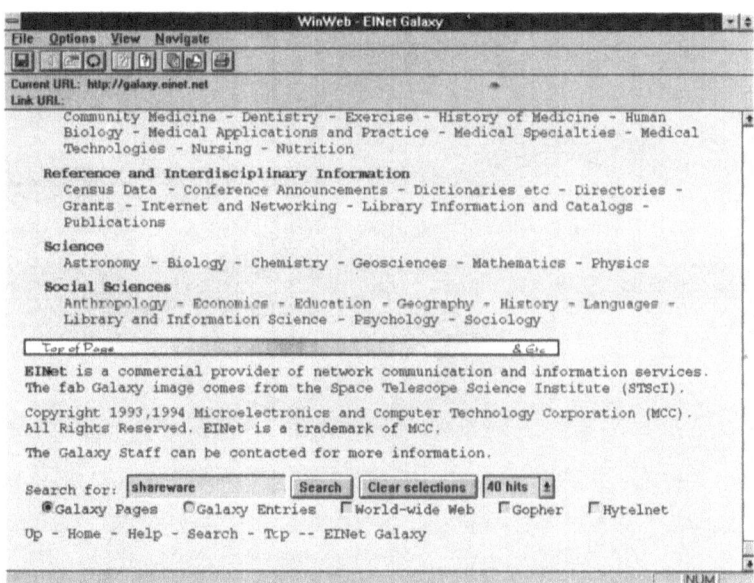

FIGURE 7.12.
The search panel at the bottom of the EINet Galaxy home page.

The results came in a hypertext list of several different types of sites. We looked for Windows software in this list, and found a number of good sites, including the Windows World shareware archive at the San Marcos campus of the California State University.

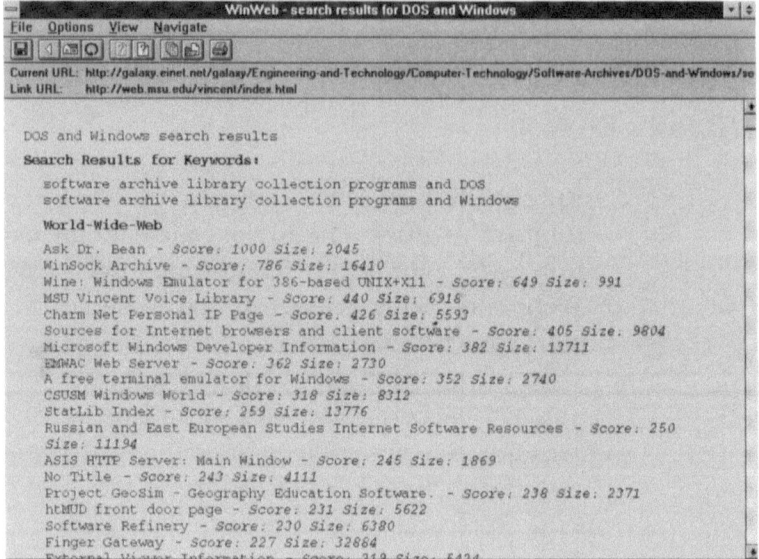

FIGURE 7.13.
Search results from
the ElNet Galaxy
home page.

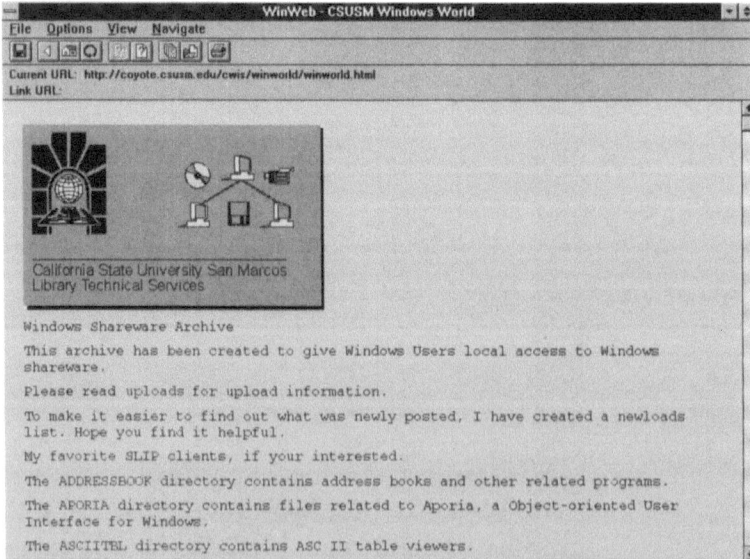

FIGURE 7.14.
The Windows World
home page at
California State
University, San
Marcos.

By following this link, we found not only a Windows share-ware archive, but one with a Web browser friendly interface (at `http://coyote.csusm.edu/cwis/winworld/winworld.html` direct-ly). It's easier to use archives that are set up for Mosaic, as they can be accessed by browsers more easily than FTP sites, and can contain more descriptive information.

This archive has been set up with subject indices for differ-ent types of programs, and is a good example of a well laid out archive.

It may take some time to sort out the different results that can be listed from a search like this, but you'll always be able to link the results to a hotlist directly, and finding a good archive is a step toward being able to use the Internet more efficiently.

8
On-Line Magazines, Journals, and Books

The Internet is a great source for online magazines, journals, and books. Several different types are obtainable, from basic text formats to interactive Web documents. Mosaic makes it easy to access all of these from its intuitive interface.

The range of topics presented is wide, ranging from scientific journals to art and architecture magazines, history journals, and literature. Publishing electronically has opened up a means of expression for a lot of people who otherwise would not have this outlet. This makes reading electronic journals a rewarding and refreshing experience. Of course, standard magazines are also taking advantage of the World Wide Web as a chance to make new interactive versions available, with interesting additions like sound and movie files not available in print versions.

To get started with electronic journals, take advantage of the many good indices established on the Internet. CERN, the Institute for Particle Physics in Switzerland, has developed the Virtual Library, a catalog of subjects linked to further hypertext collections of World Wide Web info on the particular subject. To get there from NCSA Mosaic, go to the Meta Index link listed on the home page, and select the Virtual Library from the index list. The direct URL for other Web browsers is `http://info.cern.ch/hypertext/DataSources/bySubject/Overview.html`.

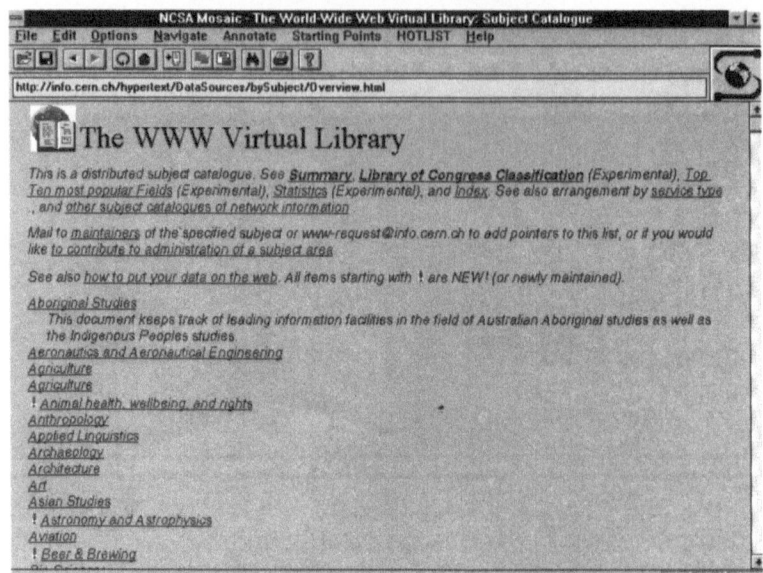

FIGURE 8.1.
The opening page for the Virtual Library index at CERN.

Go to the topic for Electronic Journals, and you'll see a further index page, with a comprehensive list of journals and magazines.

There are a number of interesting items listed, and the index is dynamic; it should continue to be updated regularly. You'll also find pointers to other electronic journal indices on the Internet here.

Continuing with the Virtual Library's list, the first thing you'll notice is that many of the journals and magazines listed are actually part of collections on Gopher servers. Gophers are a good way to index text (like on-line magazine and journals), and are easily accessible for users with slower connections.

The E-Serials gopher at CIC-Net offers a list of archived journals to browse, through a subject index and alphabetical directories. Representative magazines featured include the *Flora Online* journal from Cornell University, the *EduCom Review* (for educators using computers), and the *Electronic Journal of the Astronomical Society of the Atlantic*. Reach CIC-Net at gopher://gopher.cic.net/11/e-serials (Figure 8.4).

The Publications area at the WELL (Whole Earth 'Lectronic Link) is a public-access gopher with a wide range of interesting topics, including texts from William S. Burroughs and Bruce Sterling, electronic information from magazines like *Mondo 2000*, and specialty items like the Factsheet Five reviews of eclec-

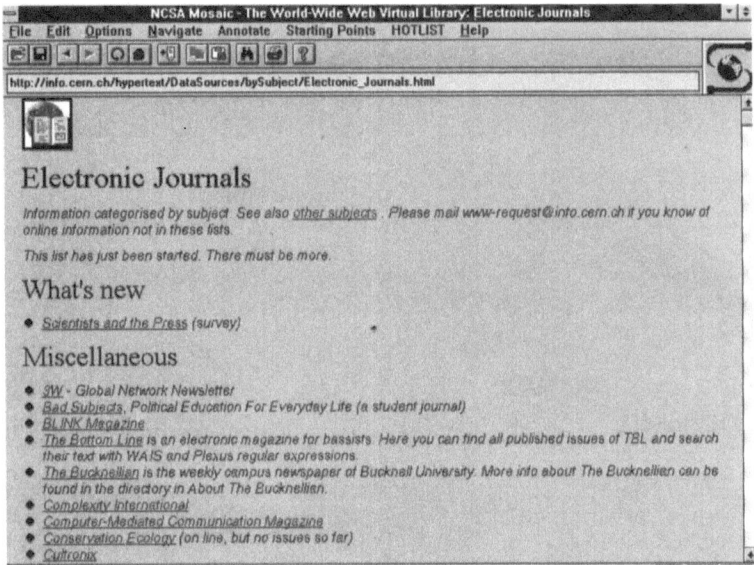

FIGURE 8.2.
Electronic Journals index at the Virtual Library, opening page.

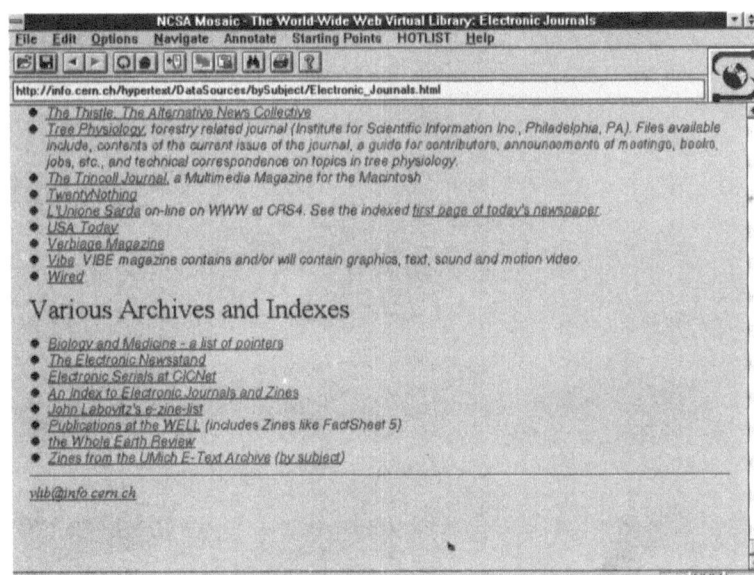

FIGURE 8.3.
Electronic Journals index at the Virtual Library, continued.

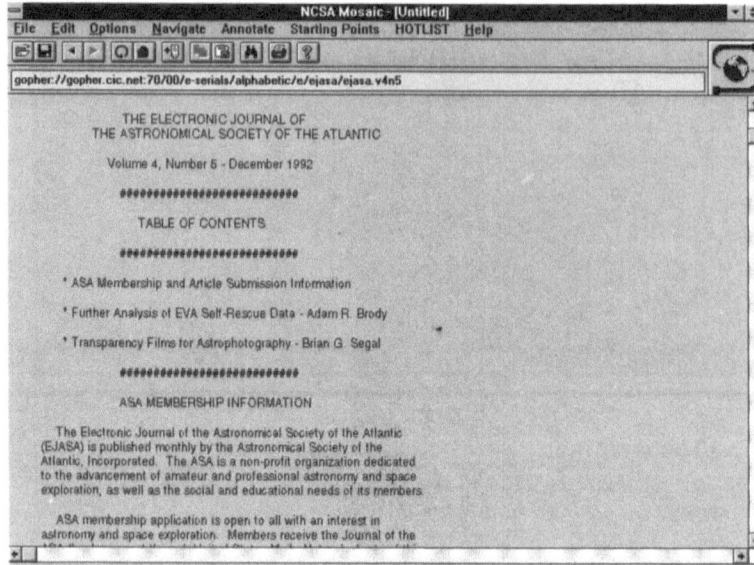

FIGURE 8.4.
The Electronic
Journal of the
Astronomical
Society of the
Atlantic at CIC-Net.

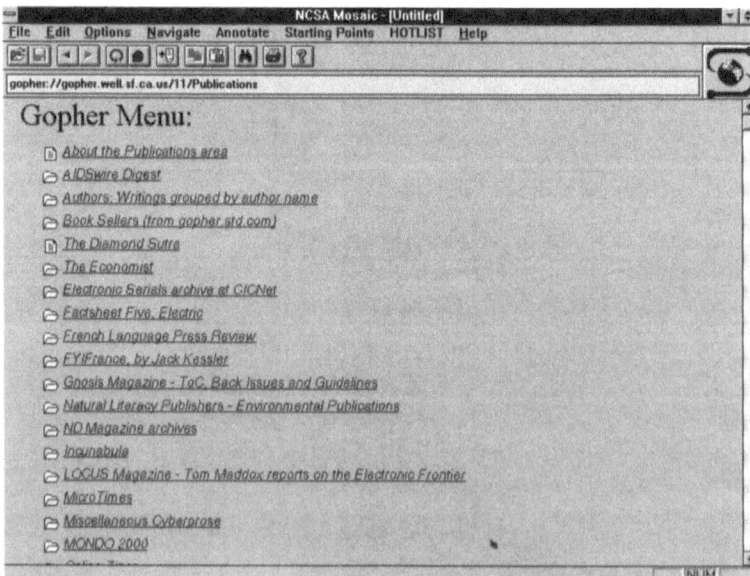

FIGURE 8.5.
The WELL's Gopher
Publications page.

tic smaller magazines and an AIDSwire Digest archive. The latter two are also searchable via a gopher search page, which makes it easy to look up topics. Use gopher to go to the WELL Publications area at `gopher://gopher.well.sf.ca.us:70/11/Publications/`.

The Electronic Newsstand, at `gopher://gopher.enews.com:2100/1`, provides mostly small samples from larger magazines, including Table of Contents and trial articles. It's not a total electronic magazine solution, but it does provide a taste of the magazines featured. Representative articles from *Alaska* magazine, the *Harvard International Review*, the *Sporting News*, and *Videomaker* show a good range of topics. Sample material from computer magazines like *CADENCE*, *Virtual Reality World*, *PC Novice*, *PC Today*, and *Computerworld*, as well as from mainstream magazines like the *New Yorker*, *Air & Space/Smithsonian*, and *Discover* rounds out the Electronic Newsstand's offerings. It's also got a searchable gopher index to its articles.

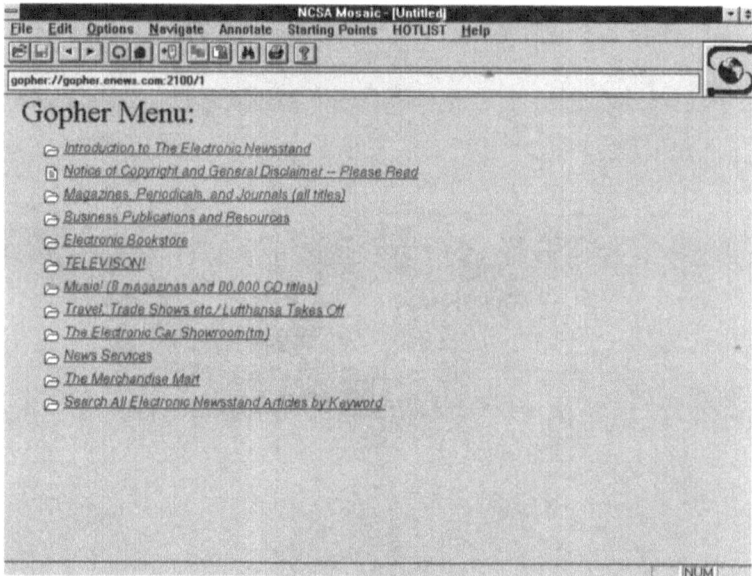

FIGURE 8.6.
The Electronic Newsstand.

Other gophers for electronic journals are located at the University of Delaware (`gopher://morris.lib.udel.edu:70`) and the University of California at Santa Cruz, which features an index of the extensive MELVYL library catalog, at `gopher://scilibx.ucsc.edu:70/11/The%20Library/Electronic%20Journals`. You'll need to have Telnet configured for Mosaic as an external program to reach some of the materials available from these gopher servers.

Beyond Gophers

Beyond the gopher interface for electronic journals, Mosaic-style hypermedia Web pages have also been developed. These are easier to navigate, using the same hyperlink interface as a home page, and can also be formatted to present their information more clearly.

You can get the same Factsheet Five information as is on the WELL's gopher server in an HTML Web browser format at Stanford University. It's at `http://kzsu.stanford.edu/uwi/f5e/f5e.html`. Note the difference between the HTML and gopher versions; the Web browser version is much easier to use with Mosaic and Netscape.

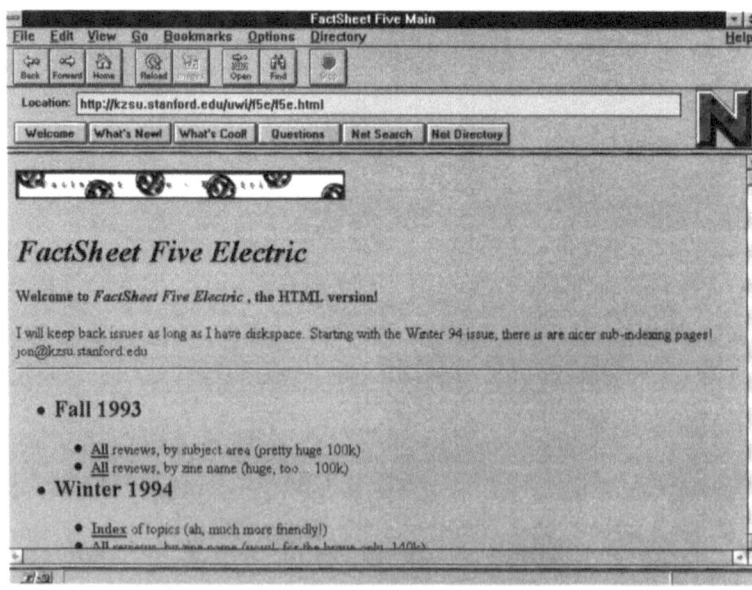

FIGURE 8.7.
The Factsheet Five Electronic page.

A good index to the biology and medical journals available on the Internet is available from Harvard University, at `http://golgi.harvard.edu/journals.html`, that includes links to peer-reviewed journals and journals extracts, as well as newspapers, newsletters, and discussion groups.

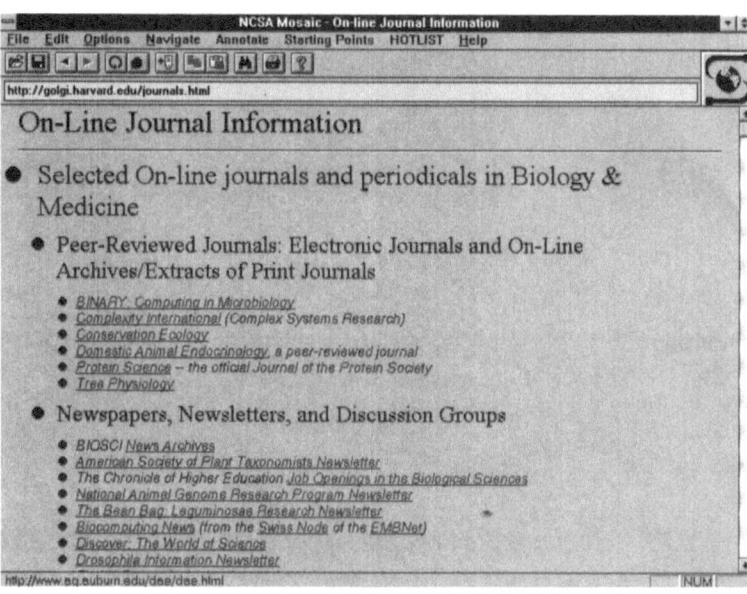

FIGURE 8.8.
The Biology and Medical Journals index at Harvard University.

Carleton College's Interesting Articles index features an in-depth look at the subject of electronic journals themselves, and links to several of them. The direct link to this site is at `http://journal.biology.carleton.ca/Journal/background/HotArticles.html` (Figure 8.9).

The Scholarly Communications Project at Virginia Tech features a number of industrial/technical electronic journals, as well as interesting articles about setting up hypermedia libraries. It's at `http://borg.lib.vt.edu` (Figure 8.10).

The DA-CLOD (Distributedly Archived Categorical List of Documents) is an index that you can add to yourself. The subject directory for literature has a number of subdirectories for books, magazines, newspapers, newsletters, and plays. Since the index is freely extensible by any individual with a compatible Web browser (like Mosaic 2.0 or Netscape), you'll find a lot of material here that wouldn't be available over the Internet otherwise. Get to DA-CLOD at `http://schiller.wustl.edu/DACLOD/daclod`.

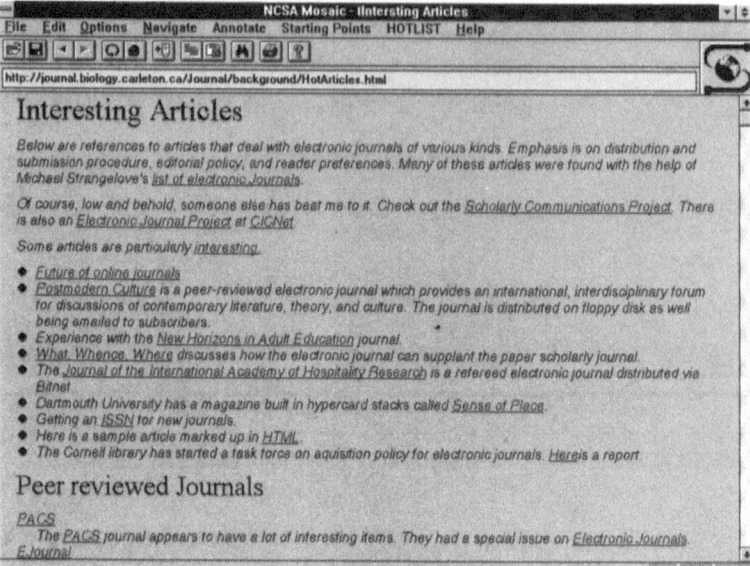

FIGURE 8.9.
The Carleton College
Interesting Articles
index.

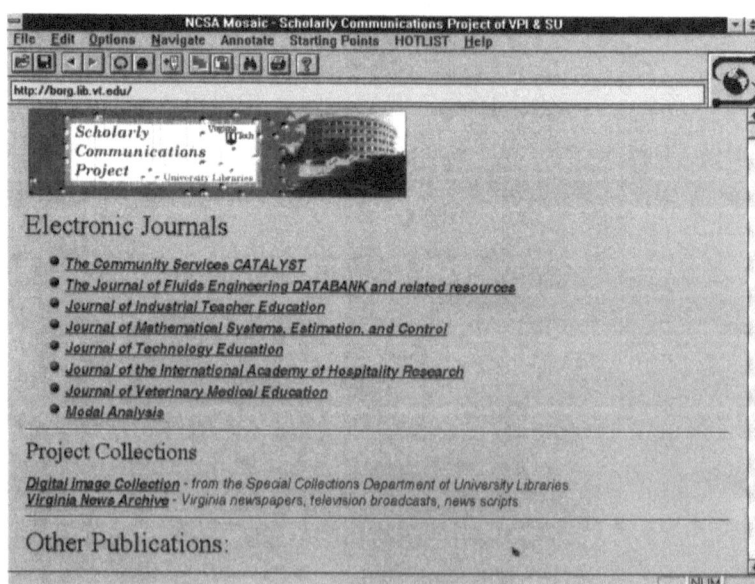

FIGURE 8.10.
The Scholarly
Communications
Project at Virginia
Tech.

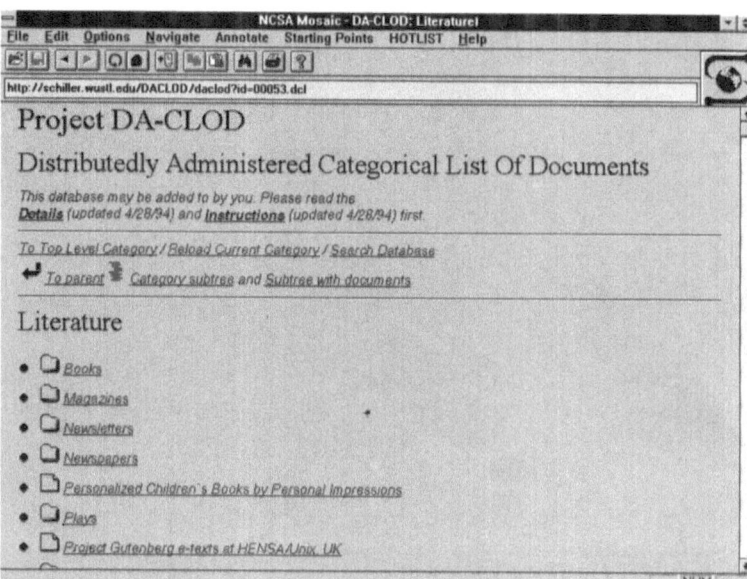

FIGURE 8.11.
Project DA-CLOD
Literature index
at Washington
University–St. Louis.

The English Server at Carnegie-Mellon University (`http://english-server.hss.cmu.edu`) is a very well laid out gateway to a large amount of information on the humanities (Figures 8.12 and 8.13). Subject areas include art and architecture journals, drama, fiction, film and TV works, and newspapers/newsletters on many different topics. You can perform Gopher searches, link to hypermedia journals, and even leave mail for the editors at the site.

Individual Web Browser Magazines and Books

One of the most interesting natural developments following the creation of Mosaic-style Web browsers has been the generation of interactive magazines and books. These can include stylized graphics, animations, and sound, as well as links to other Web sites and interactive elements like feedback via email. There are cutting-edge periodicals like avant-garde student productions, as well as major mainstream publications like *Mother Jones* and *Time* magazine, in a rich hypermedia formats.

Cultronix is a part of the Carnegie-Mellon University World Wide Web project. It's a cultural studies journal, and is a good

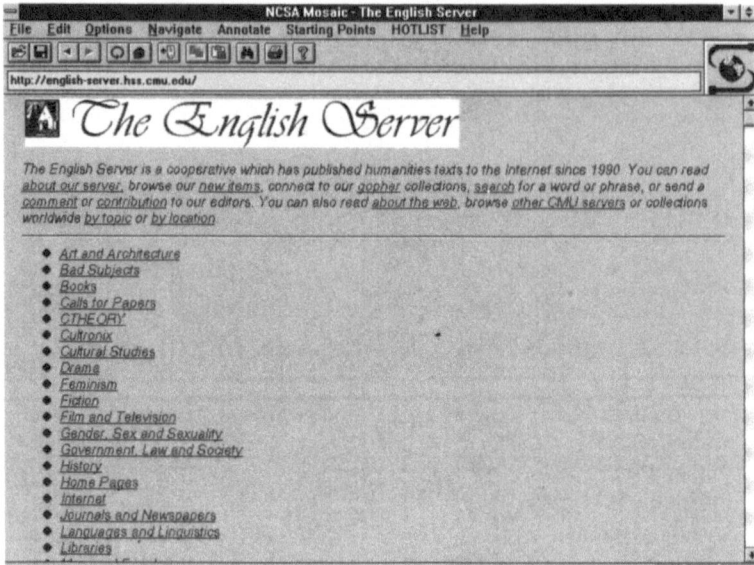

FIGURE 8.12.
The English Server at Carnegie-Mellon University.

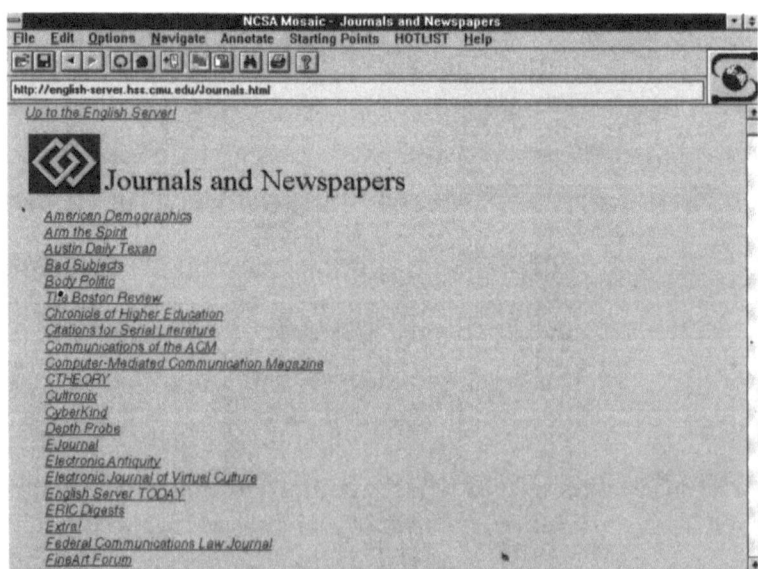

FIGURE 8.13.
The English Server Journals and Newspapers index.

embodiment of a World Wide Web interactive magazine. It offers text, graphics, sound and video, and is located at `http://english-server.hss.cmu.edu/cultronix.hmtl`.

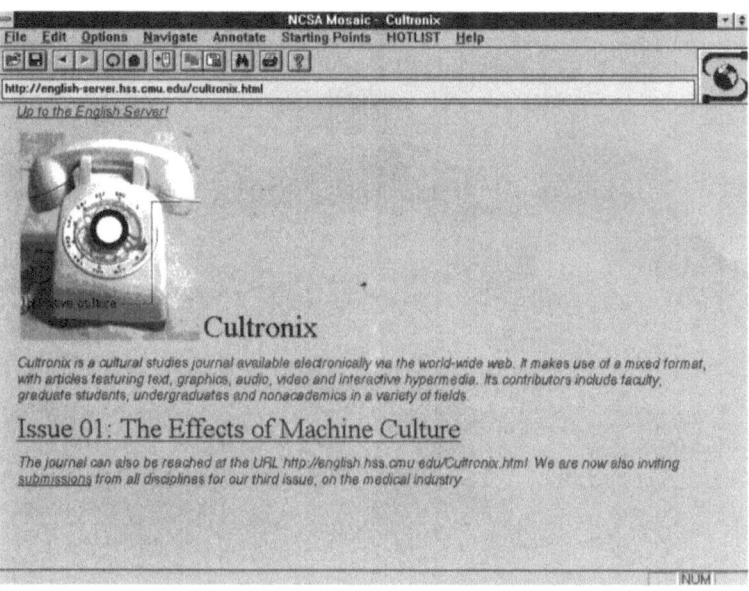

FIGURE 8.14.
Cultronix cultural studies journal at Carnegie-Mellon University.

The *CyberZine* is a collection of graphics-oriented counterculture multimedia offerings, with an emphasis on using the World Wide Web for communication. The summer 1994 issue also has links to information about the Internet, including the Enterprise Integration Technologies' "Guide to Cyberspace" (also available from `http://www.eit.com:80/web/www.guide/guide.toc.html`). Catch a sample issue of CyberZine at `http://www.cyberzine.org/html/CyberLearn/Summer94/summer.html` (Figure 8.15).

Departure From Normal is a fun art journal that shows the capabilities of the Web to good effect. One of its recent issues featured a QuickTime movie in addition to text-based works. You can get there directly at `http://www.teleport.com/~xwinds/dfn.html` (Figure 8.16).

Verbiage, despite its name, is a good collection of original fiction available on-line. It's edited by Thomas Boutell at the University of North Carolina, and it seems to be a stable place for original literature on the World Wide Web. Check it out at `http://sunsite.unc.edu/boutell/verbiage` (Figure 8.17).

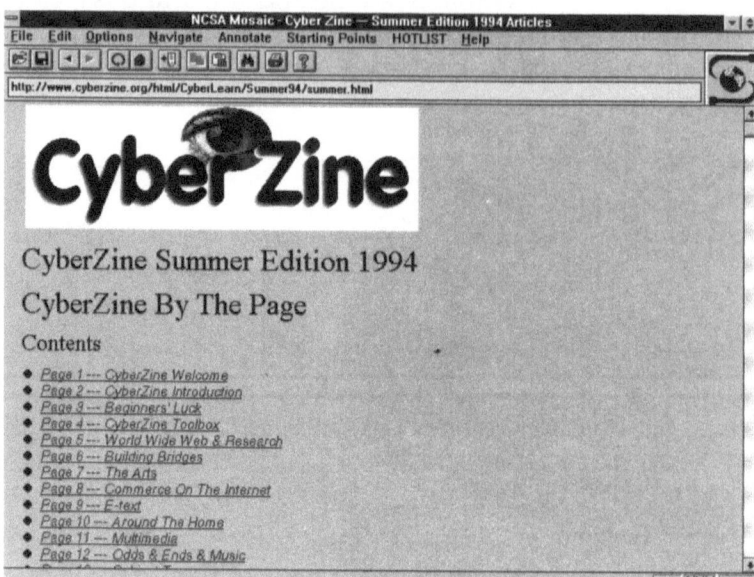

FIGURE 8.15.
CyberZine magazine opening page.

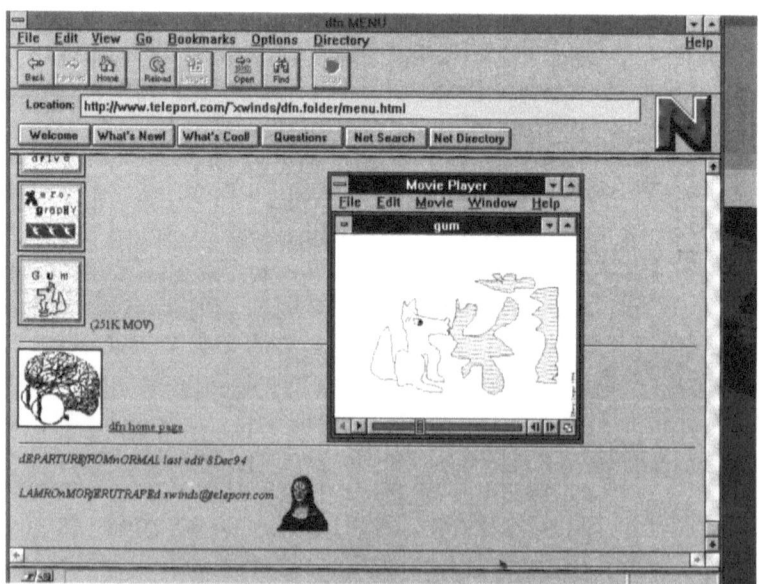

FIGURE 8.16.
Departure From Normal Table of Contents, with external QuickTime movie playing.

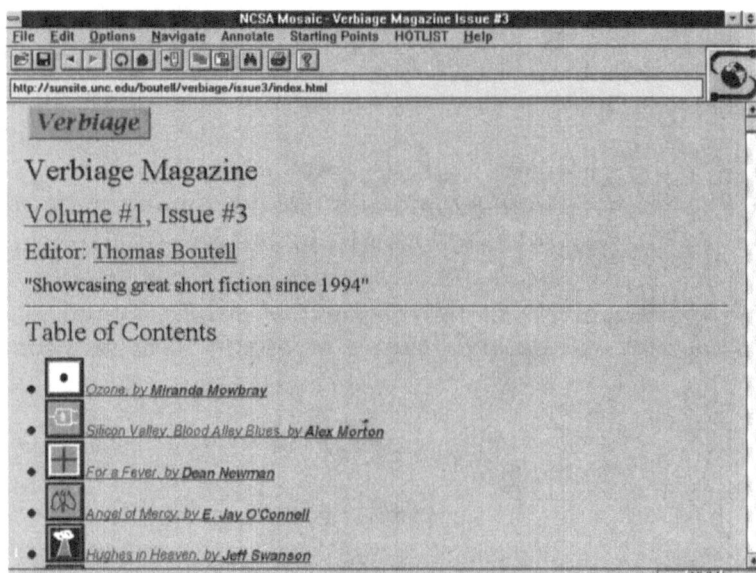

FIGURE 8.17.
Verbiage fiction magazine.

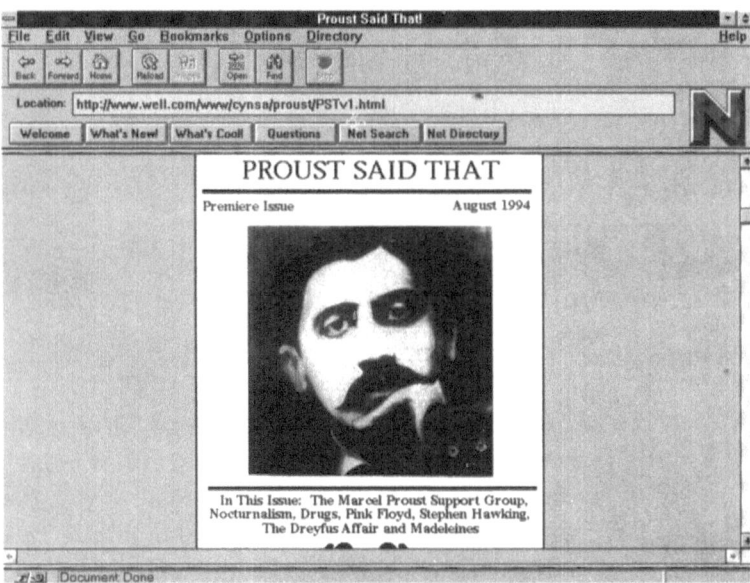

FIGURE 8.18.
Proust Said That!
magazine.

The smaller eclectic magazines are also worth checking out, if you can find them. The *Proust Said That!* on-line magazine, at `http://www.well.com/www/cynsa/proust/PSTv1.html`, is an in-depth look at the 19th-century French writer. It includes short stories, background information, and even recipes (Figure 8.18).

The larger-scale magazine efforts on the Web include a full hypermedia version of *Mother Jones*, *Wired Magazine's HotWired Web* site, and *Time Interactive*. These are all experimental versions of on-line magazines, with World Wide Web interfaces and varying degrees of sophistication.

Mother Jones's HTML version is straightforward and clean, and shows up well under Mosaic. It includes full-text articles with graphics, on-line classifieds, and a special section called *Mother Jones Interactive*, a fully digital meeting place to learn about and discuss sociopolitical issues. *Mother Jones* on the Web is at http://www.mojones.com.

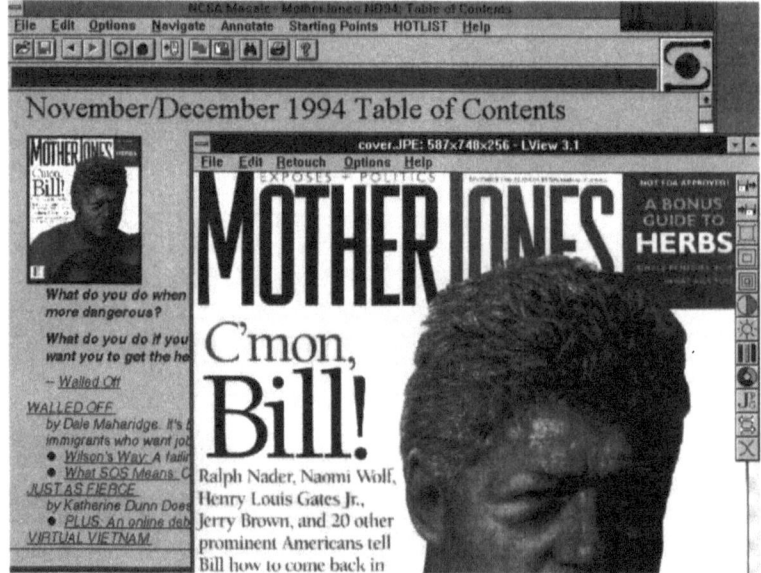

FIGURE 8.19.
An example on-line issue of *Mother Jones* magazine.

HotWired, *Wired* magazine's reworking of parts of their magazine into a digital format, is a hit-or-miss proposition. It's not exactly intuitive, too cute to be useful, and frantically in love with its own overstyling. On the other hand, there are good articles with graphics, sound, and animation files located here, and if you can stand the advertising (of a kind you'll find almost nowhere else on the Web), you may find something interesting. Try it yourself at http://www.hotwired.com. You'll have to use Netscape, not NCSA Mosaic, to access *HotWired*, due to what's claimed to be a bug in NCSA Mosaic's interface.

The vast Time/Warner Internet site called Pathfinder is a good

FIGURE 8.20.
HotWired's main
menu, with several
QuickTime movies
available from the
Renaissance 2.0
section.

place to find slick magazines on the World Wide Web, including *Time* features and back issues, *Entertainment Weekly* reviews, and *Money* magazine advisories. Other divisions like Time Warner Publishing are also represented. The offerings are well-suited to the Web, including JPEG stills in the magazines and movie reviews and QuickTime previews of upcoming books and CD-ROMs. The Pathfinder system also includes messaging areas, like electronic bulletin boards. This is an opportunity to use the Time Web site like NetNews, though it's for local access. The Time universe is located at `http://www.timeinc.com/time/universe. html` (Figure 8.21).

Playboy Enterprises has an on-line version of *Playboy* magazine available at `http://mosaic.playboy.com`. The December 1994 issue featured an interview with Bill Gates of Microsoft, not to mention the usual advice columns and downloadable Playmate photographs.

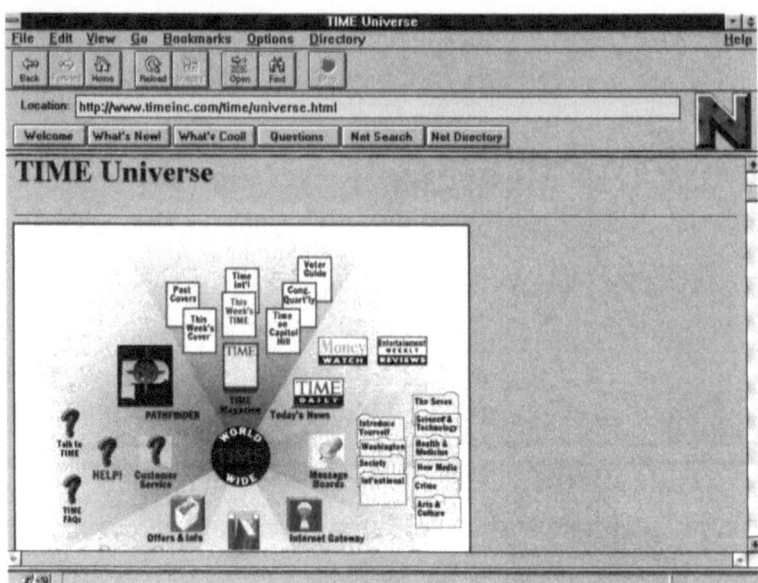

FIGURE 8.21.
The *Time* magazine
Web site overview.

Books, Literature, and Miscellaneous

GNN Magazine, a World Wide Web resource, is in its own category. The Global Network Navigator from O'Reilly and Associates is an attempt to provide a networked magazine to the Internet. It's the home page for Internet in a Box, and provides an electronic copy of *The Whole Internet Catalog* that's useful for navigating the Internet. It also features a Best of the Net section and online magazines on topics like personal finance and travel. GNN is more heavily connected to the Internet than most online magazines, and has many useful links to a wide array of interesting sites. Get there from Internet in a Box, or by going to http://gnn.com/gnn/GNNhome.html.

The *Daily Telegraph*

The *Electronic Daily Telegraph,* an on-line version of the British newspaper, is a fine example of a graphics-intensive online newspaper. It features up-to-date news reports and editorial features, as well as full-color newswire photographs. For free registration, use the http://www.telegraph.co.uk address.

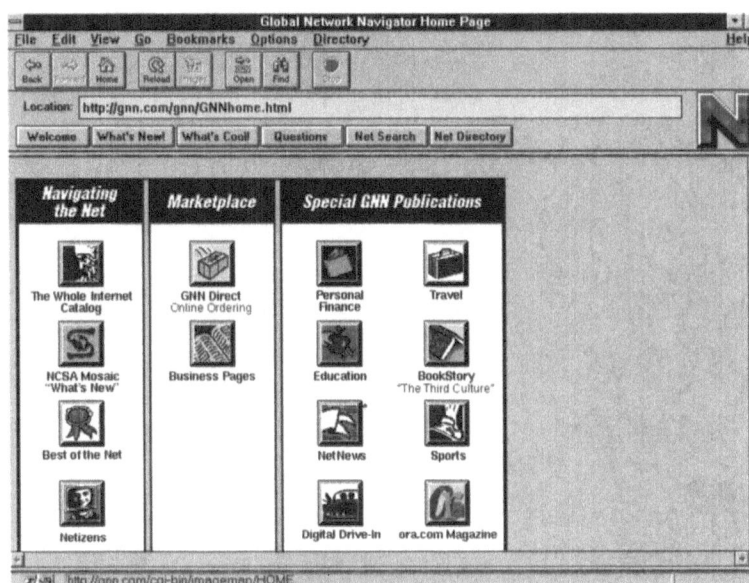

FIGURE 8.22. Representative offerings from O'Reilly and Associates' Global Network Navigator.

Project Gutenberg

Project Gutenberg is an ambitious distributed Web project designed to put electronic books on-line. These range from basic downloadable text files to hypermedia documents that you can view with Mosaic. A representative example is an electronic version of Lewis Carroll's *Alice In Wonderland*, from Carnegie-Mellon University, with original illustrations by Sir John Tenniel (available at `http://www.cs.cmu.edu:8001/Web/People/rgs/alice-table.html`) (Figure 8.23).

Other HTML books available include fiction, featuring a lot of Edgar Rice Burroughs, Bram Stoker's *Dracula*, Mary Shelley's *Frankenstein*, Mark Twain's *The Adventures of Tom Sawyer*, and Melville's *Moby Dick*, among others. Essays and nonfiction includes Thomas Paine's *The Age of Reason*, the works of John Donne and St. John of the Cross, the collected inaugural addresses of the U.S. Presidents, Sun Tzu's *The Art Of War*, and books on programming languages and book conservation. Poetry collections range from *Leaves of Grass* to the complete works of Wordsworth, and also include works by John Keats, Oscar Wilde, and Samuel Taylor Coleridge. Modern writing includes Hakin Bey's *Temporary Autonomous Zone* (poetic terrorism) and Philip Greenspun's *Travels With Samantha* (with color photos).

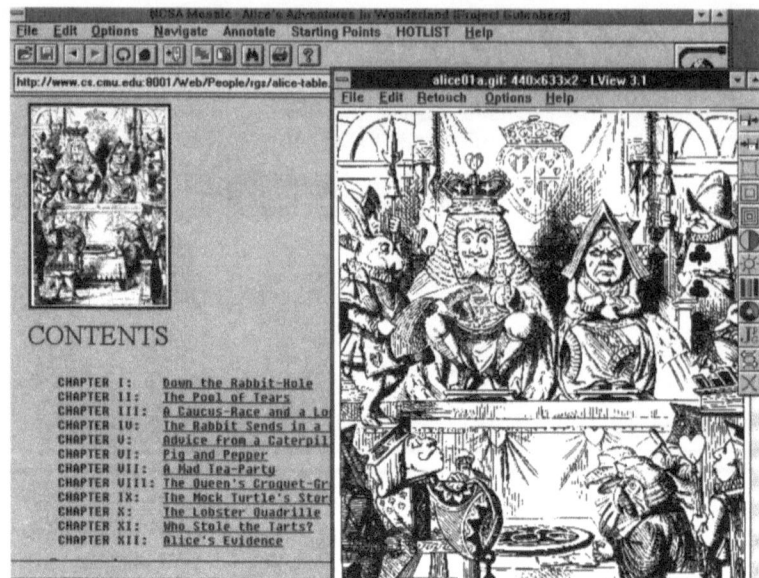

FIGURE 8.23.
The illustrated *Alice In Wonderland* on the World Wide Web.

You can find a short list of HTML Mosaic-readable books at `http://www.cs.cmu.edu:8001/Web/People/rgs/`. Pointers to comprehensive lists of on-line works are located at the Project Gutenberg home page (`http://med-amsa.bu.edu/Gutenberg/Welcome.html`). You can also go directly to an alphabetical index at `http://med-amsa.bu.edu/Gutenberg/alpha.html`.

The Library of Congress Web site, located at `http://lcweb.loc.gov/homepage/lchp.html`, is a must-see for people interested in a good on-line library source. It includes special exhibits that run under Mosaic Web browsers (for example, a recent selection included African-American culture, the voyage of Columbus, a tour of the Vatican library, and a view of the Dead Sea scrolls), select handbooks to worldwide countries, and links to the MARVEL and LOCIS on-line catalog systems. There are also sample pages for a proposed Global Electronic Library, with links to state, local, and federal World Wide Web sites, and a good collection of Internet resources and searchable indices.

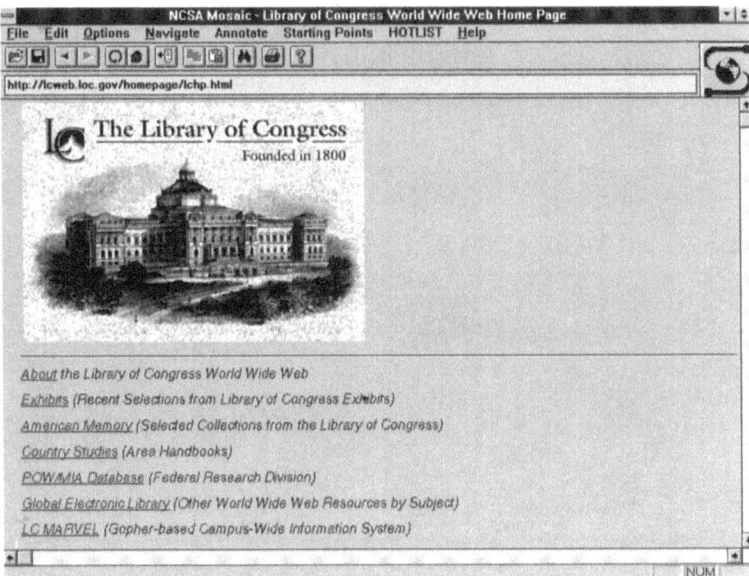

FIGURE 8.24.
The Library of
Congress Web site.

Finally, it's worth checking out the Literature index at the Virtual Library at CERN. It's the home of the Internet Book Information Center, a good link to information on literature all around the Internet. The IBIC home page includes links to collected Web sites for authors, booksellers, publishers, libraries, and online reference tools.

9
Art, Games, Music, and More

The World Wide Web is the home of many different types of artistic, intellectual, and fun pursuits, including online art galleries, comics-related home pages, music sites, and interactive games. There are also many different indices to this information available in formats that you can easily navigate in your Mosaic Web browser.

Art and Art Galleries

The art featured on the World Wide Web can be different from what you're used to seeing. It's a bit more freewheeling, loose, and inventive. The Mosaic interface is used to good effect in these interactive types of artworks, however, so it's definitely worth a look. A good starting point is the University of Kentucky Art-Source index (`http://www.uky.edu/Artsource/artsourcehome.html`), which provides a comprehensive index to the art world (Figure 9.1).

You can also try the ArtROM online museum guide, located at `http://www.primenet.com/art-rom/museumweb/`, as a way to find art museums on the Web. These range from city and county local institutions to large international sites.

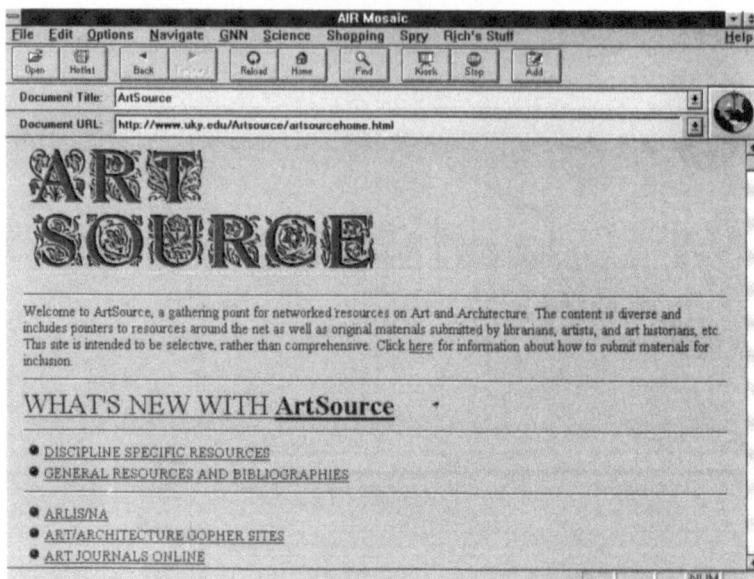

FIGURE 9.1.
The ArtSource home page.

One museum to definitely visit on the Web is the Louvre, available in a Mosaic browser tour from `http://mistral.enst.fr/` (select the Sunsite mirror at the University of North Carolina for faster access from the United States: `http://sunsite.unc.edu/louvre/`). This site has a very nice collection of images from the museum, including classic paintings, French medieval artwork, and a tour of the city of Paris. It also includes classical music sound files and video clips from French television.

The Virtual Library's Museums listings (at `http://www.comlab.ox.ac.uk/archive/other/museums.html`) are also worth browsing to find museums of many different types on the Internet. Try the Art listings as well (`http://info.cern.ch/hypertext/DataSources/bySubject/Literature/Overview.html`).

Galleries on the Web range from one-man shows like the Strange Interactions Art Show, an exhibition of paintings by John Jacobsen (`http://amanda.physics.wisc.edu/show.html`), and the truly haunting work *Life With Father* (`http://gertrude.art.uiuc.edu/ludgate/the/place/stories/life_with_father/Life_With_Father.html`), by Joseph Squier, to collections of works like the WERDNA Cruciform Art Gallery (`http://www.ugcs.caltech.edu/~werdna/`).

Other gallery collections include the School of Visual Arts in

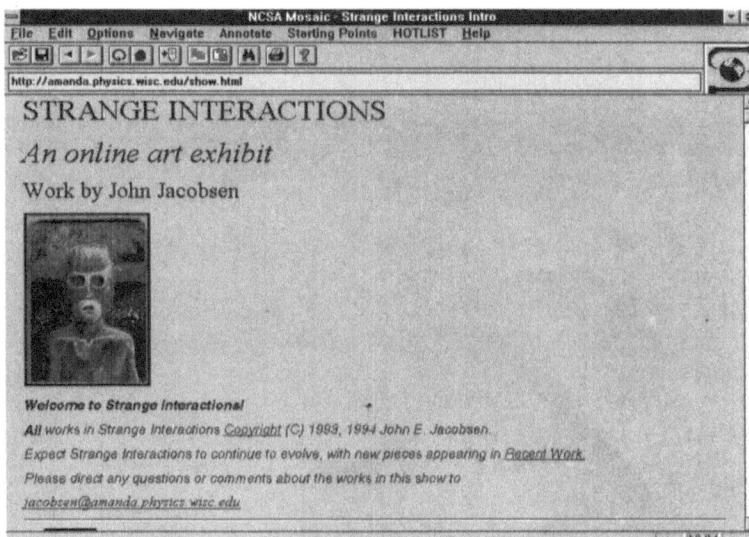

FIGURE 9.2.
The Strange
Interactions art
show, paintings by
John Jacobsen.

New York's "Who's Got the Body," a collaborative exhibit consisting of web sites maintained by artists linked to the show by a digital map (located at `http://www.sva.edu/WGTB/flypaper.html`), and the (art)^n Laboratory at Northwestern University (`http://www.artn.nwu.edu/`), featuring a virtual photography exhibit.

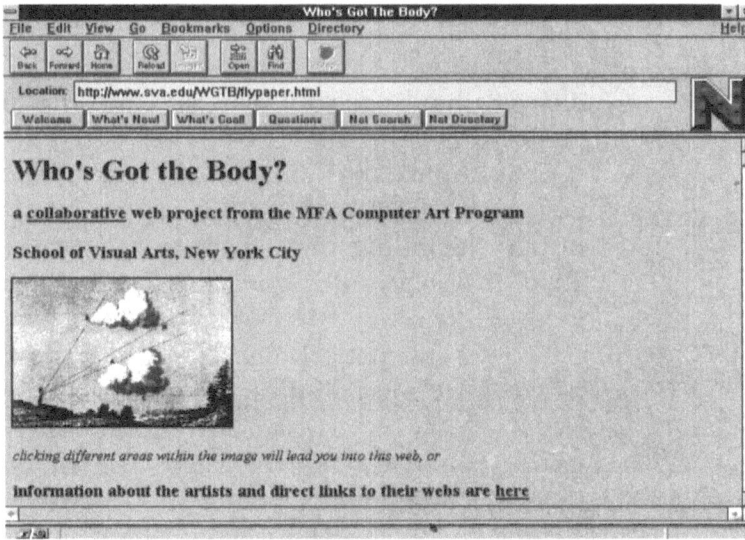

FIGURE 9.3.
"Who's Got the
Body?" Exhibit at
the School of Visual
Arts in New York.

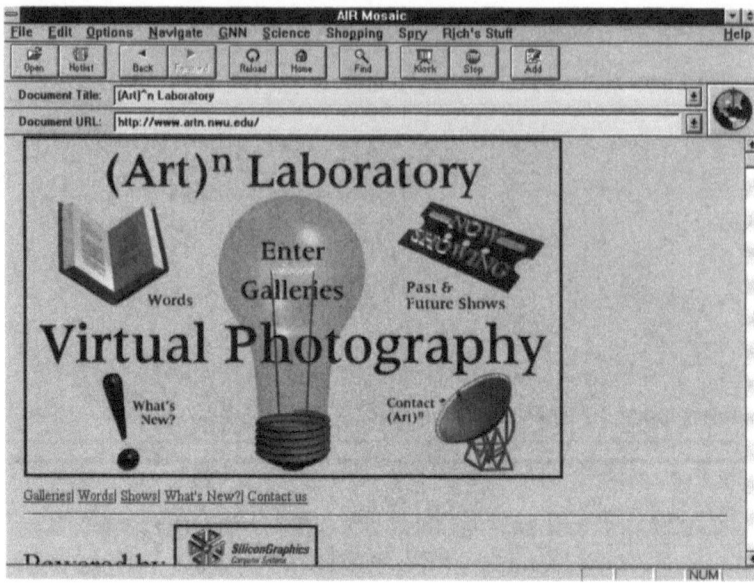

FIGURE 9.4.
The (art)^n
Laboratory Virtual
Photography exhibit.

Art exhibits and galleries on the World Wide Web change over time, so be sure to check the indices for current shows.

Alternative Home Pages

The Internet is populated by people with a sense of humor, and an appreciation for the arts. It's a rough sort of humor sometimes, and may not be to your taste. On the other hand, the experience can be fun.

For example, the Roadside Attractions page (`http://reaction.scruznet.com/ROADSIDE.HTML`) is a good collection of off-the-wall humor, including the adventures of Marnie, an online comic strip. It also features good links to a collection of weird spots located all around the World Wide Web.

You can also check out The Asylum and the Creative Internet pages at the California Institute of Technology, `http://www.galcit.caltech.edu/~ta/cgi-bin/asylhome-ta` and `http:www.galcit.caltech.edu/~ta/creative.html`, respectively. These feature interesting items like interactive artworks, including a Lite Brite box you can add to yourself, as well as an interactive TV index, a Websurfer guide, and an Internet polling booth.

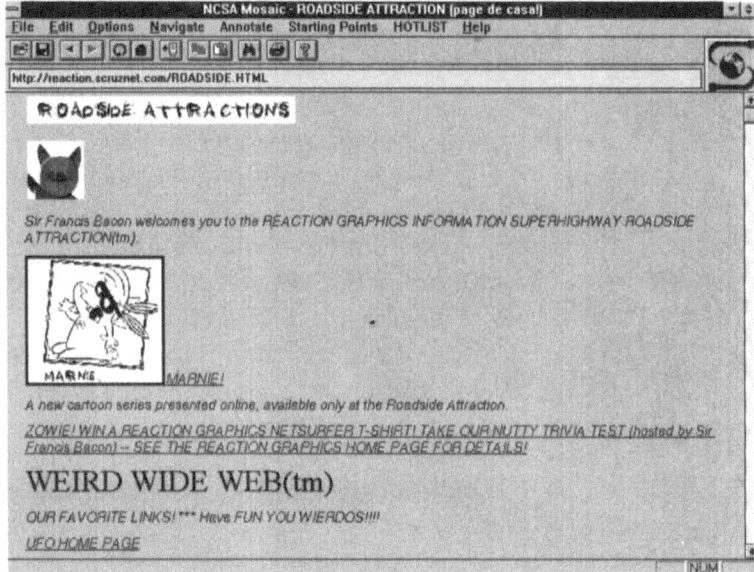

FIGURE 9.5.
The Roadside
Attractions home
page.

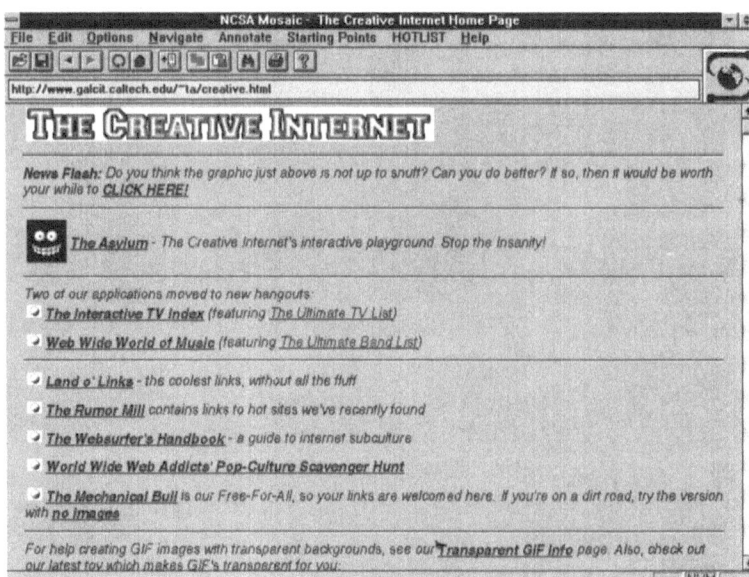

FIGURE 9.6.
The Creative
Internet page at
CalTech.

CyberSight's home page (`http://cybersight.com/cgi-bin/cs/s?main.gmml`) is also worth taking a look at. It includes an interactive grafitti wall that you can embellish with your own graphics, hypermedia exhibits of different types, and links to interesting sites on the Web.

For the type of art you can't find anywhere but on the Web, try an interactive art exhibit focusing on genetics located at `http://porsche.boltz.cs.cmu:8001/htbin/mjwgenform` (yet another project from Carnegie-Mellon University). This site allows you to interactively create fractal representations of genetic structures by entering in your own parameters into a form in your Mosaic page. You can then view the artwork you've created in your Web browser or in an external viewer. You can also vote on your creations at the site.

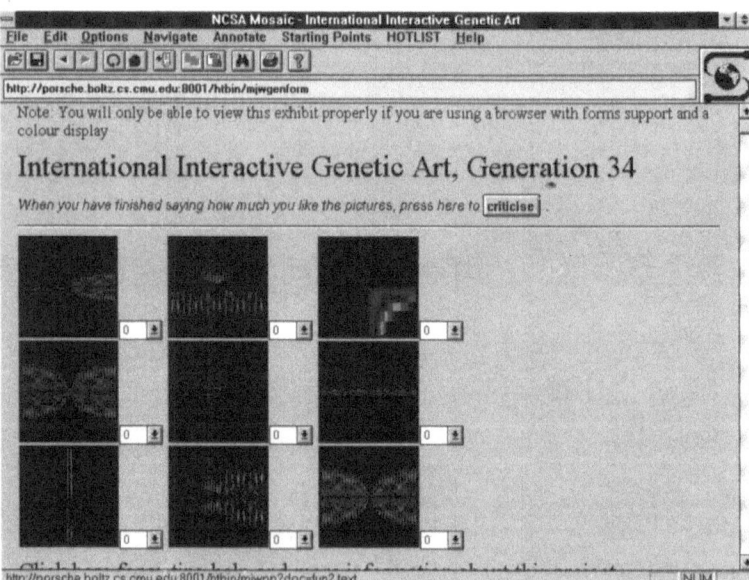

FIGURE 9.7.
The International Interactive Genetic Art project.

The Syracuse University Computer Graphics for the Arts home page (`http://ziris.syr.edu/home.html`) is another good site for collaborative Internet art exhibits of many types. It shows off the capabilities of Mosaic for this type of interactive artwork to good effect.

The Geometry Center at The University of Minnesota features a Web site with interactive on-line geometry exhibits, including in-

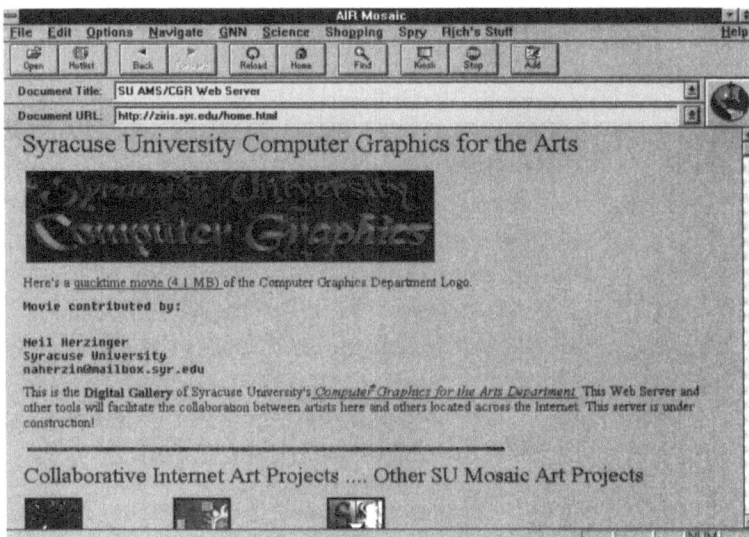

FIGURE 9.8.
The Syracuse
University Computer
Graphics for the Arts
home page.

teractive models and game prototypes that you can explore with a Web browser like Netscape. It's at `http://www.geom.umn.edu/apps/gallery.html`.

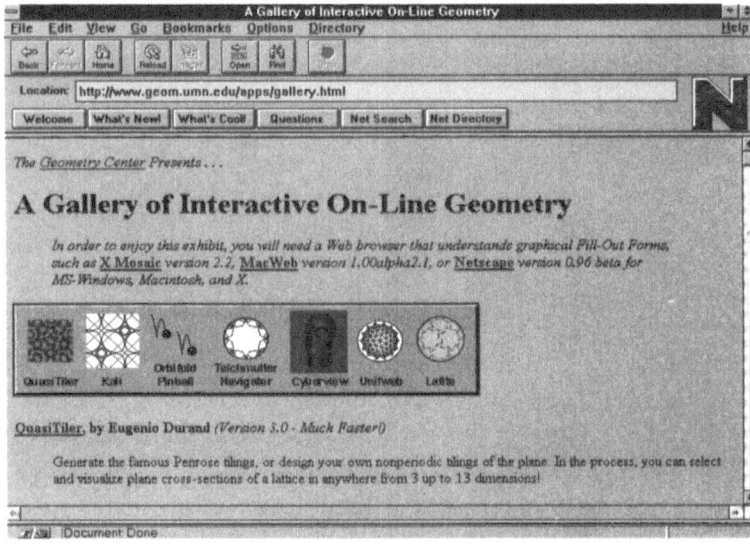

FIGURE 9.9.
The Geometry
Center at the
University of
Minnesota.

Comics on the Web

There's a lot of good information on comics on the Web. The best places to look for information are a couple of well-maintained comics-related home pages. Johann's Comics Page at `http://www.cen.uiuc.edu/~jb2561/comic.html` has links to comic strips, comics-related home pages, reviews, and more. It's well laid out and leads to many interesting places.

We went from Johann's page to a comics archive at the University of Michigan that includes cover art from a variety of different comics types and eras. It provides a good look at how comics can be viewed across the Web. It's located at `http://www.css.itd.umich.edu/users/kens/comics.html`.

FIGURE 9.10.
The Comics archive at the University of Michigan.

Another site, the Comic Book and Comic-Strip page (which calls itself "the nexus of all comic book web pages"), is located at `http://dragon.acadiau.ca:1667/~86099w/comics/comics.html`. It's also a good jumping-off point for comics-related information.

Individual artists are also featured on the World Wide Web, ranging from George Herriman's "Krazy Kat" (`http://dragon.acadiau.ca:1667/~860099w/comics/krazy_kat.html`, or via a link at the above site), to Scott Adams's "Dilbert" comic strip (`http://nearnet.gnn.com/gnn/news/comix/dilbert.html`), and Tom Tomorrow's political satire "This Modern World" (`http://www.well.com/Community/comic/`).

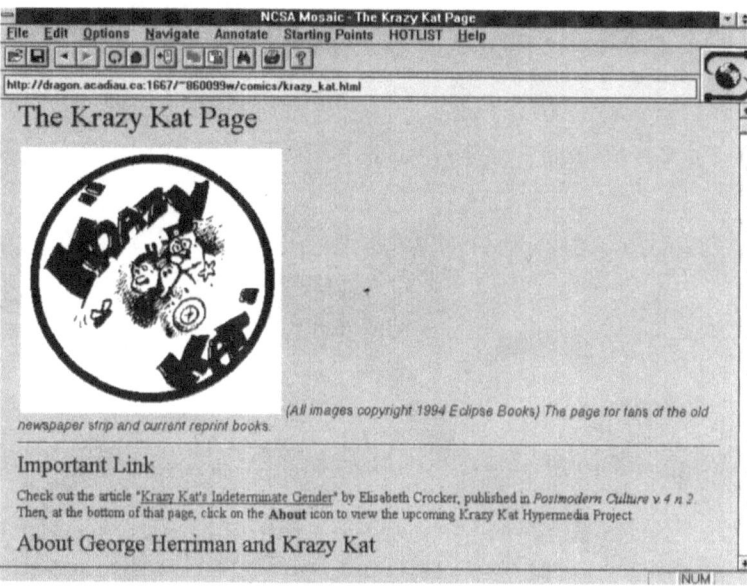

FIGURE 9.11.
The "Krazy Kat"
home page.

For an interesting look at how comics and the Web can interact, the Oncolink Web server at the University of Pennsylvania has excerpts from the Harvey Pekar work "Our Cancer Year," a realistic portrait of the artist's struggle with cancer. The link for this is at `http://cancer.med.upenn.edu:80/0h/psycho_stuff/comics/pekar_1`.

Games and Related Materials

Game information on the Internet ranges from home pages for different game systems to interactive games built right into the Web. As with the comics-related pages, the best places to start looking for information are on index pages maintained by individuals with a keen interest in gaming.

Michel Buffa's videogames page at Carnegie-Mellon University (`http://www.cs.cmu.edu:8001/afs/cs.cmu.edu/user/buffa/www/videogames.html`) is a good example. It includes links to information on game systems ranging from the SuperNintendo to the Atari Jaguar, the 3DO, and arcade machines. There's also a lot of

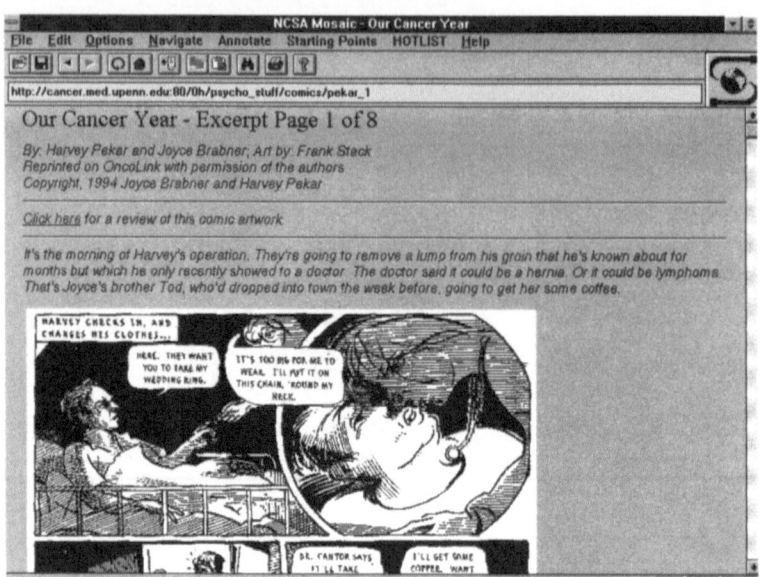

FIGURE 9.12.
"Our Cancer Year"
excerpt at OncoLink.

links to the popular multiplatform game DOOM, and a hypertext version of the videogames FAQs (Frequently Asked Questions).

SEGA of America is one of the first game system companies to have their own home page on the Net. It features information on SEGA games and game platforms, JPEG images, and MPEG movies from current and projected games, and game tips and tactics. Catch SEGA at `http://www.segaoa.com`.

For a look at a really cool game company, check out Rocket Science Games. Their site also features stills and MPEG movies from current and upcoming games. Rocket Science is at `http://www.rocketsci.com`.

Zarf's Interactive Games List is where you'll find links to most of the Net-specific games on the Internet. This site is very well laid out, and provides a good icon-based menu that lets you know the platform requirements for the different games. Games playable directly from Mosaic and other Web browsers include Tarot and I Ching fortune telling, Backgammon, Tic-Tac-Toe, Minesweeper, and more. There are also links to interactive quizzes and art projects. The Zarf list, also at Carnegie-Mellon University, is located at `http://www.cs.cmu.edu:8001/afs/cs.cmu.edu/user/zarf/www/games.html` (Figure 9.15).

Games Domain (`http://wcl-rs.bham.ac.uk/~djh/index.html`), located on a server in the United Kingdom, is another site

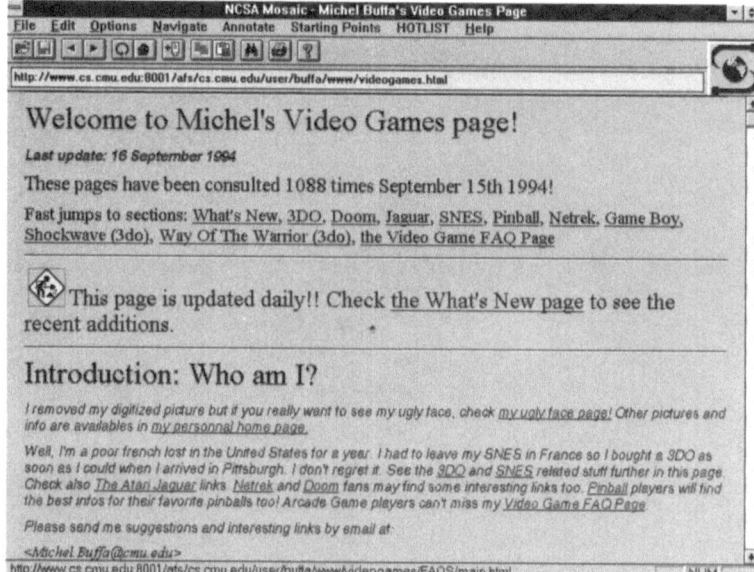

FIGURE 9.13.
Michel Buffa's
Videogames home
page.

FIGURE 9.14.
Rocket Science
Games on the Web,
with an example
JPEG image and
MPEG movie.

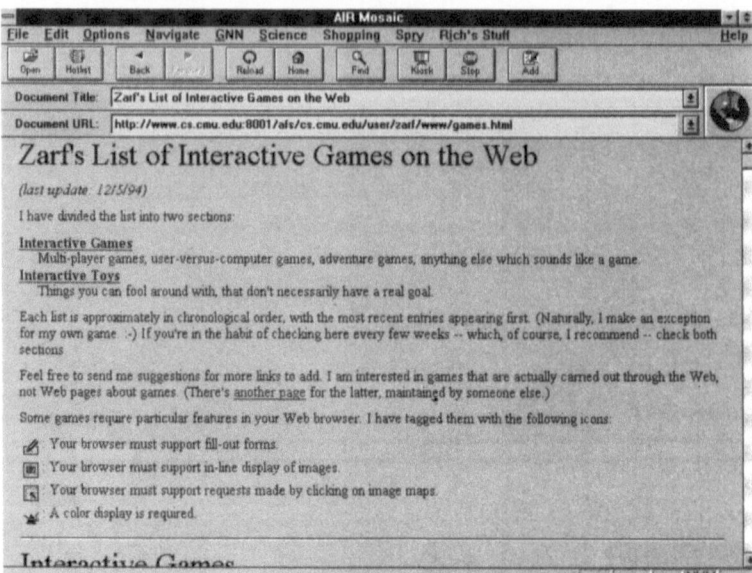

FIGURE 9.15.
The Zarf Interactive
Games home page.

with interesting links to games-related materials.

If you're looking for shareware games for your DOS/Windows system, there's a lot that is available over the Internet. The World Internet Technologies home page puts a good Mosaic interface to the larger FTP games directories. WIT is located at `http://www.wit.com/`.

Film and Animation

Film and animation sites on the Internet are very well represented, ranging from Web sites maintained by major movie studios to computer animation collections from colleges and universities. Viacom/Paramount (try `http://here.viacom.com`) and MCA/Universal (`http://www.mca.com/`) offer everything from on-line opening night participations to promotional material such as stills, MPEG video clips, and interactive multimedia software kits from current movie releases.

The deeper you get into films, the more you'll appreciate the excellent resources of the Cardiff Movie Database Browser. This site is a wealth of information on films, film directors, screenwriters, stars and their different roles, and production crews. It's set up almost exactly like a Veronica search engine, but with a cus-

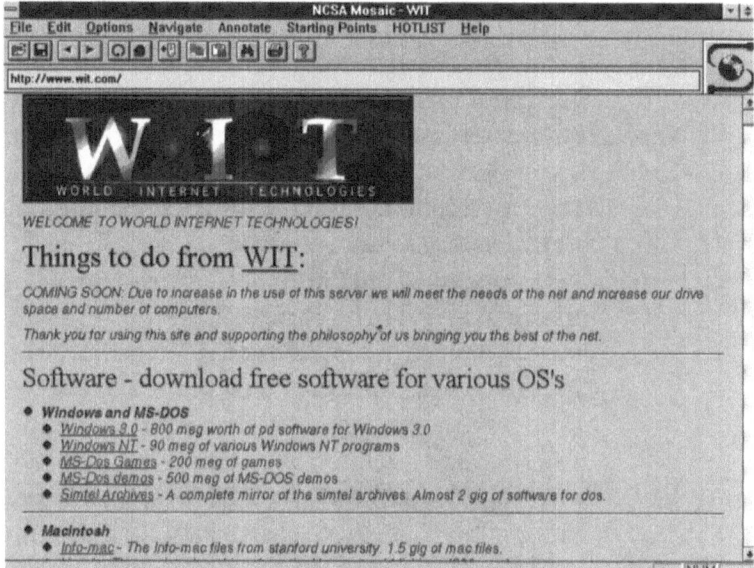

FIGURE 9.16.
The World Internet Technologies home page.

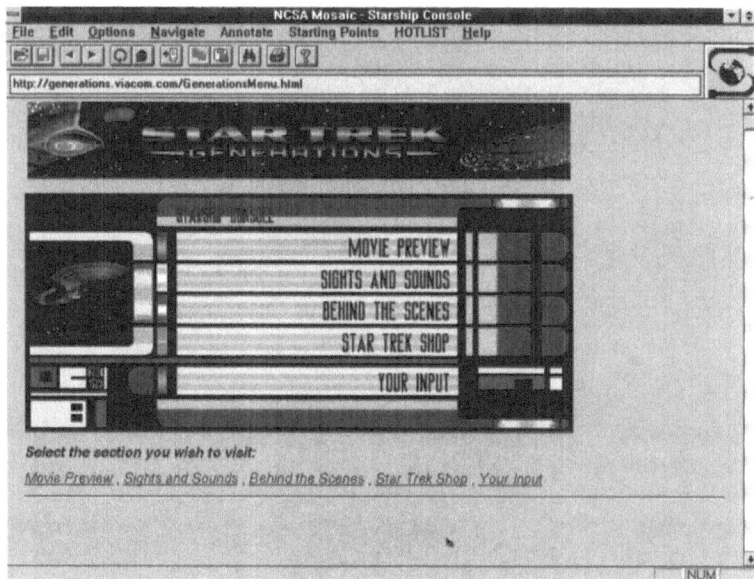

FIGURE 9.17.
Viacom/Paramount's Star Trek: Generations home page.

tom Mosaic interface that makes searching the movie database very intuitive. You can look up actors by name, and films by partial title, for example. It's easy to find a particular actor's roles or look up everything by a particular screenwriter in this database. The Cardiff movie browser is very popular, so it's been replicated at several servers around the world for more immediate access. Users in the United States should use the site at the University of Mississippi. The home page for it is located at `http://www.msstate.edu/Movies/`.

Computer animation is featured in a number of interesting sites on the Web, including Cyberia, at the Hyperreal site (`http://www.hyperreal.com/cyberia`). Cyberia's Web site is directly related to their computer animation cable TV show, and it includes Web links to sites around the world featuring animation archives.

For comics-related animation, look at Rei's page at MIT (`http://www.mit.edu:8001/people/rei/home.html`). This site features comprehensive links to the world of Japanese animation (animae), including other home pages, file archives, episode guides, and fan magazines.

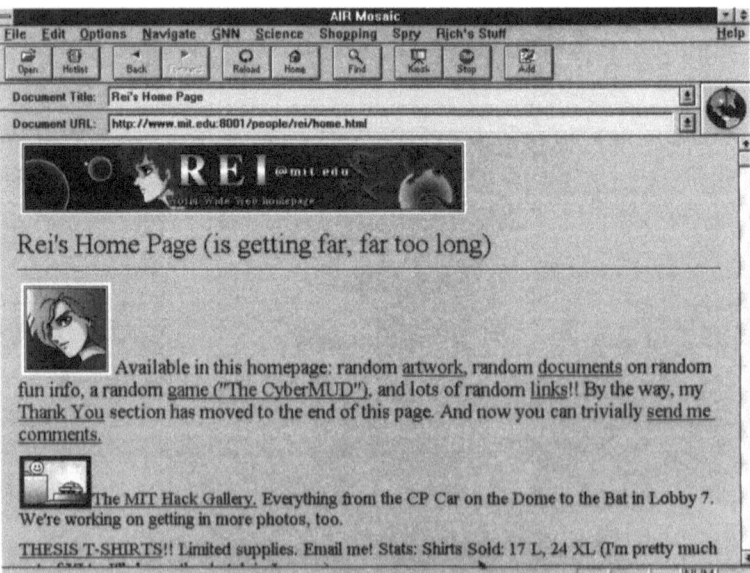

FIGURE 9.18. Rei's home page at MIT, focusing on Japanese animation (animae).

Music Sites

Some very comprehensive guides to music sites on the Internet can be found at one or two sites, including the Internet Underground Music Archive (http://www.iuma.com/) at Stanford University, and the Web Wide World Of Music (http://american.recordings.com/wwwofmusic/index.html), maintained by American Recordings, Inc. Both feature links to bands all around the world, and home pages that feature album cover art, digital music samples, and concert information. The Web Wide World site also has a link to the Ultimate Bands list, an alphabetical index to home pages for various music artists all across the Internet. You can also link your favorite band's site, if it's not already listed.

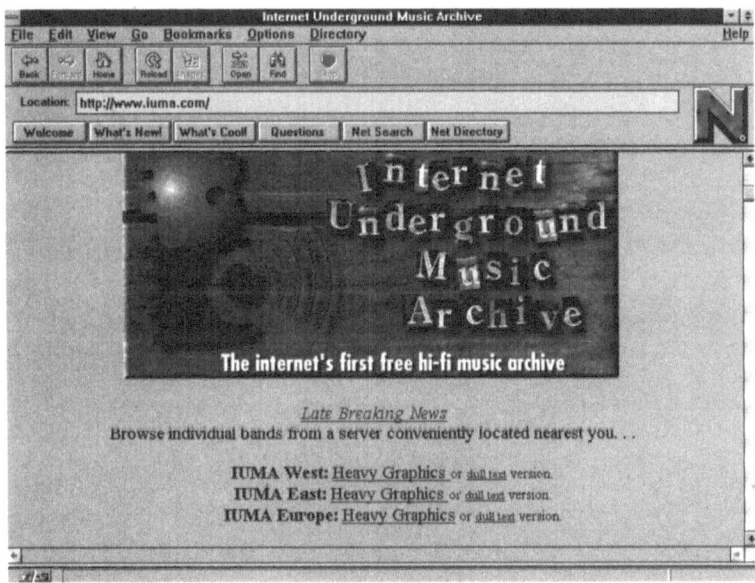

FIGURE 9.19.
The Internet Underground Music Archive at Stanford University.

Other music sites you may want to try include EmeraldNet (http://www.Emerald.NET/), keepers of the Machines of Loving Grace home page, the Virtual Radio archive (at http://www.microserve.net/vradio/), and Motown Records (at http://www.musicbase.co.uk/music/motown/).

This is just a sample of the music resources available over the Internet; you'll also want to look at the Virtual Library's Music section, located on a server in Finland at http://www.oulu.fi/music.html, and also the music subcategory of the Whole Internet Catalog Arts department at GNN (http://nearnet.gnn.com /gnn/wic/art.toc.html). The latter also includes links to former MTV VJ Adam Curry's interesting Metaverse Internet site, and to home pages for artists like Elvis Presley and Frank Zappa.

Miscellaneous Unclassifiable Fun and Toys

Practically unclassifiable are some of the more interesting Internet sites. The Lego page is one of these; it includes information on building kits, instruction sheets, and contest information. It also includes specialty items like reports on tours of the Lego factory and on MIT's Lego robots courses. Find it at http://legowww.homepages.com.

The Froggy page at Yale University (http://www.cs.yale. edu/HTML/YALE/CS/HyPlans/loosemore-sandra/froggy.html) is your key to bufo things worldwide, including great graphics and a link to an interactive frog dissection kit that runs under Mosaic Web browsers.

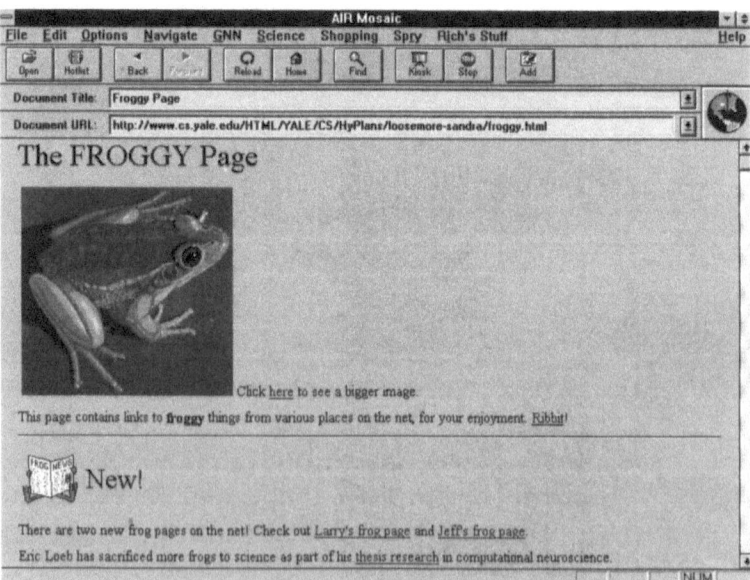

FIGURE 9.20.
The Froggy page at Yale University.

A pair of interesting interactive sites are the Trojan Room Coffee Machine at the Cambridge University (http://www.cl.cam. ac.uk/coffee/coffee.html) and the Amazing FishCam at Netscape Communications (`http://www.mcom.com/fishcam/index.html`). These take photographs of their respective subjects at around-the-clock intervals and put the photos up on their servers. At any time, you can check on the status of a coffeepot in England or on a fish tank in Northern Calfornia. Remember, if you can't see anything, the lights may be turned out....

The Klingon Language Institute provides a World Wide Web interface to their projects, which include creating and maintaining a dictionary of the Star Trek alien culture's language and translating selected works into Klingon. Find the institute at `http://www.kli.org/`.

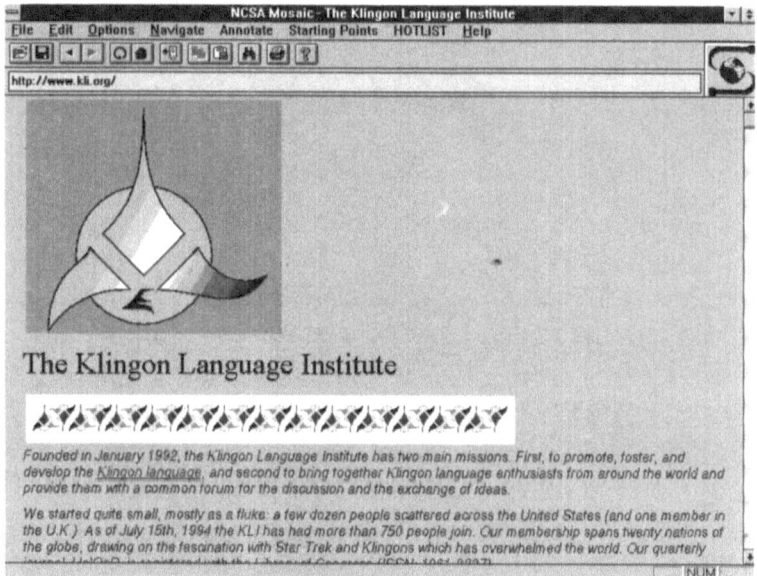

FIGURE 9.21.
The Klingon
Language Institute.

One of the more interesting fringe Internet projects is a World Wide Web guide based on Douglas Adams's *Hitchiker's Guide to the Galaxy*. This is a whimsical collections of links articles about the Internet and Internet culture, and is a good antidote to hard-core Net surfing. The Project Galactic guide is located at Willamette University, at `http://www.willamette.edu/pgg/articles.html`. It also includes a search index, and you're encouraged to submit your own articles.

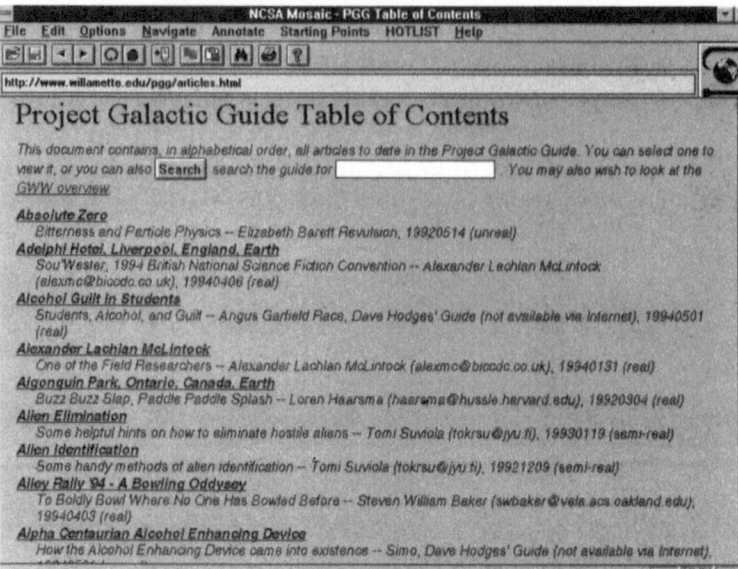

FIGURE 9.22.
The Project Galactic
Guide.

10
Serious Productivity

If you're looking for Internet resources on a more serious note, you can find a wealth of information on business, including investment risk analysis and stock information, as well as a vast collection of weather-related information, and also several interesting Web servers being run by government agencies.

Business Centers and Information Services

The Internet has several good centers for business. A central site is at the University of Texas server maintained by Professor James Garven. The RISKWeb at U. Texas (`http://riskweb.bus.utexas.edu:80/riskweb.html`) is aimed at providing World Wide Web insurance risk information, and includes links to several important sites, including on-line documentation and statistics. Its FINWeb counterpart (`http://riskweb.bus.utexas.edu/finweb.html`) is a Financial Economics server with more general links to finance issues (Figure 10.1).

The University of Texas site also includes a helpful page of links to commercial sites on the Web, at `http://riskweb.bus.utexas.edu:80/commerce.html`. These include Web site collections like CommerceNet (focusing on commercial Internet sites),

FIGURE 10.1.
FINWeb at the
University of Texas.

at http://www.commerce.net/, and IndustryNet (focusing on the manufacturing industry), at http://www.industry.net/, as well as alternate indices for commerce on the Net, like Thomas Ho's huge collection of links at http://biomed.nus.sg/people/commmenu.html.

Also on the University of Texas server, the Kiwi Club (http://kiwiclub.bus.utexas.edu/finance/kiwiserver/kiwiserver.html) maintains a comprehensive Web site with many good links to financial lecture information, bank and corporate sites, and commercial companies on the World Wide Web.

The Clearinghouse for Subject-Oriented Resources project at CERN also features a Mosaic guide to business sites, issues, and information on the Internet. You'll find it at a server at Louisiana State University, at http://www.lib.lsu.edu/bus/index.html. It provides a good overview of business resources on the Net in a hypertext subject guide (Figure 10.4).

GNN's Whole Internet Catalog business listings (http://gnn.com/gnn/wic/bus.toc.html) include subtopic links to Web sites on agriculture, career and employment services, entrepreneurship and small business information, Internet commerce, investment and management houses, nonprofit organizations, personal finance, and real estate matters.

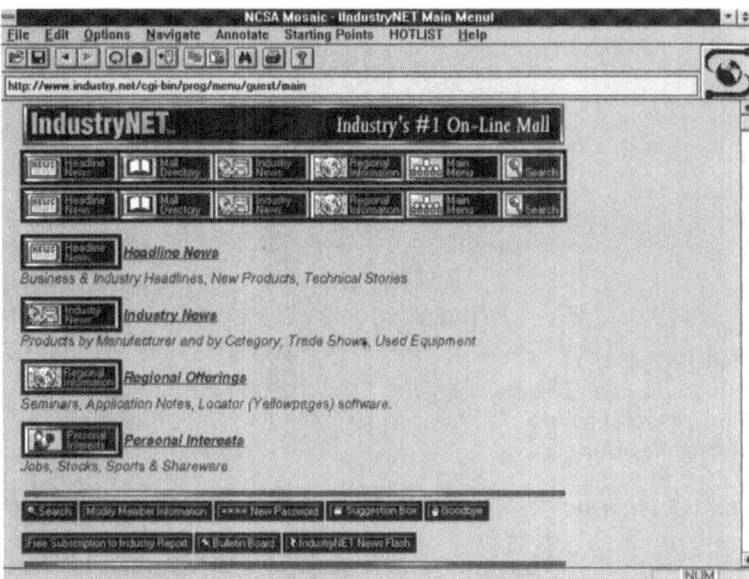

FIGURE 10.2.
IndustryNet home
page.

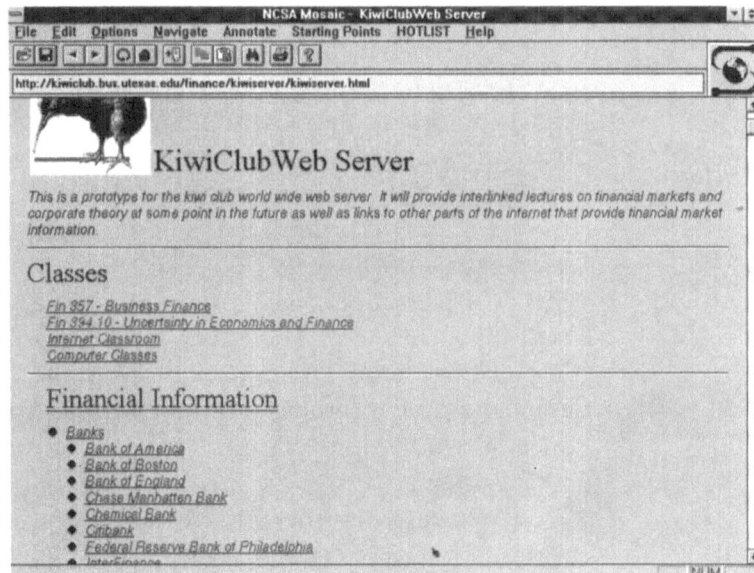

FIGURE 10.3.
The Kiwi Club Web
site at the University
of Texas.

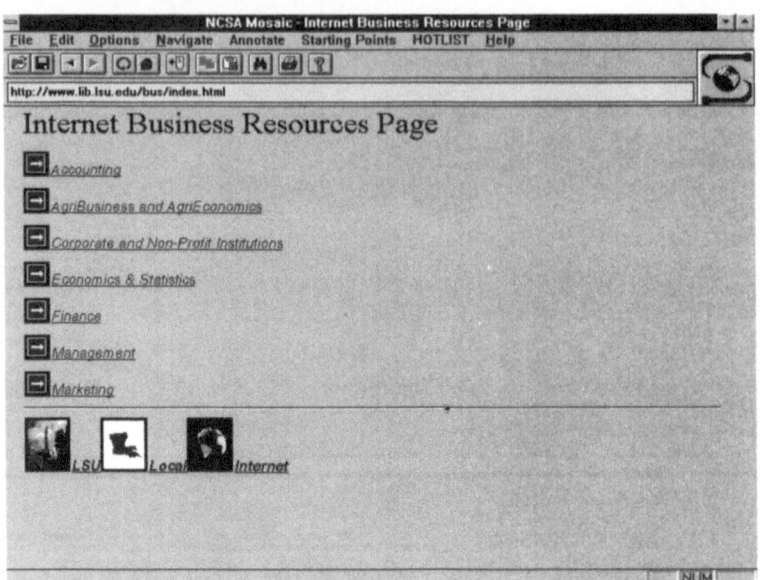

FIGURE 10.4.
The Clearinghouse for Subject-Oriented Resources Business Guide at Louisiana State University.

Also at GNN, the Personal Finance Center (`http://nearnet.gnn.com/gnn/meta/finance/index.html`) provides a collection of resources for the private investor, including articles on how to invest wisely and links to stock quote retrieval systems and mutual fund information.

Netscape Communications' What's Cool page has a link to the Rensselaer Polytechnic Institute short list of interesting business sites on the World Wide Web (`http://www.rpi.edu/~okeefe/business.html`); this features a prescreened list of no more than 50 businesses deemed worthy of inclusion to the list because of superior presentation, suitability of the Web as a medium for their business, and other criteria. It's a good place to find interesting Web sites directly.

Also in Netscape's list is the Internet Business Directory, at `http://ibd.ar.com/`. The IBD is a subscription Web site for commercial companies. Commercial Web site collections often include searchable indices, company directories, as well as home pages for individual companies with contact information and special offers.

You can also try the on-line Guide to Commercial Services on the Net (`http://www.directory.net/`), as well as the Internet Business Center (`http://www.tig.com/IBC/index.html`), and

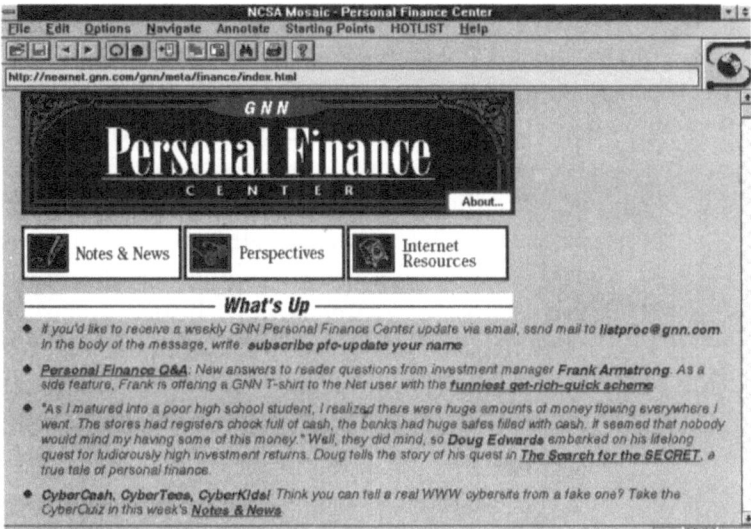

FIGURE 10.5.
The GNN Personal
Finance Center.

GNN's Commercial Business listings (`http://gnn.com/gnn/bus/index.html`), for more information on commercial corporations on the Net.

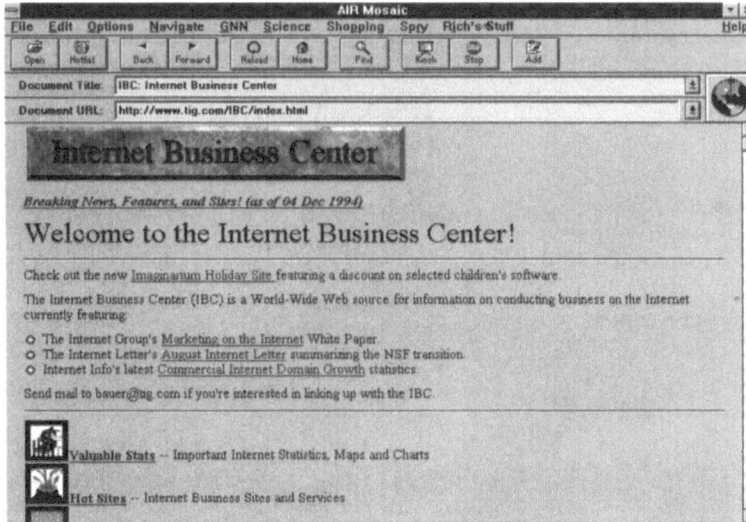

FIGURE 10.6.
The Internet
Business Center.

Banks and Investment Houses

There are many banks and investment houses on the Internet, and you can use an index like the Kiwi Club at the University of Texas listed above to access them. Two representative examples include the Wells Fargo Bank Web site (http://www.wellsfargo.com/), which features an alphabetical index to the bank's services and multimedia sound and picture files of the bank's history, and the J.P. Morgan investment house (http://www.jpmorgan.com/index.html), which also features an index to Morgan's services on the Web, as well as a link to an interactive Web application called RiskMetrics (used for market risk management).

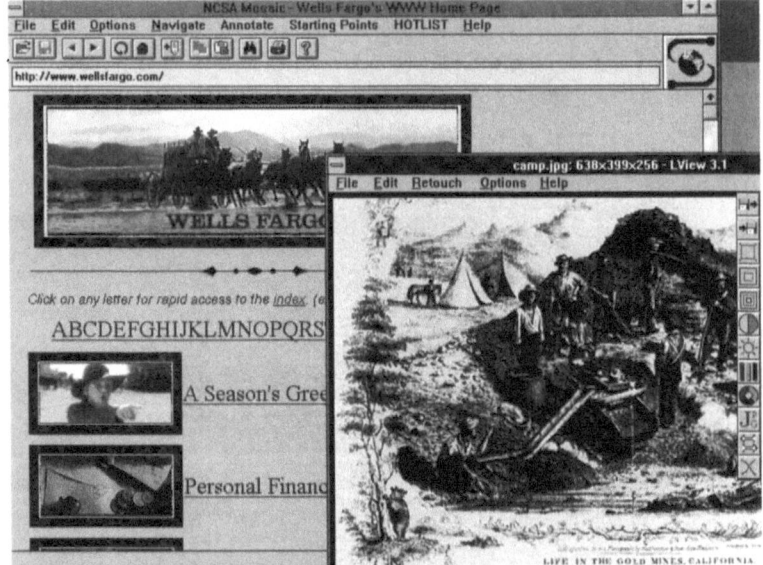

FIGURE 10.7.
The Wells Fargo Bank Web site (with an example JPEG image from their historical archive).

You may also find the Chicago Mercantile Exchange Web site useful, at http://www.interaccess.com:80/users/wilbirk/.

FIGURE 10.8.
The J.P. Morgan investment house home page.

Government Web Sites

The University of Texas RISKWeb site also includes a page with a Government World Wide Web server list (http://riskweb.bus.utexas.edu/levin/govern.html), which includes links to the Fed-World Federal information system, as well as home pages for the White House, the FBI, the Department of Commerce, and the Social Security Administration. You can find further government information at the U.S. Patent and Trademark Office site (http://www.uspto.gov/) and the Small Business Administration On-line service (http://www.sbaonline.sba.gov/) (Figure 10.9).

Utilities and Services

Business services include the DowVision information retrieval system of articles from *The Wall Street Journal* (http://dowvision.wais.net/ for a trial version), and Security APL's stock quote interface, which lets you look up time-delayed market quotes by entering in the appropriate market symbols. The PAWWS (Portfolio Accounting World Wide–Security APL) home page is

FIGURE 10.9.
The Small Business
Administration
On-line.

located at `http://pawws.secapl.com/`, and it also features information like SEC filings. You can get to the quote server directly at `http://www.secapl.com/cgi-bin/qs`.

The NetWorth Personal Investment service page, a subscription service with information on over 5000 mutual funds, is located at `http://networth.galt.com/`.

The Maxwell Labs Taxing Times server provides online tax information, including electronic forms available for downloading. It's located at `http://www.scubed.com:8001/tax/tax.html`.

You can check on Federal Express packages by entering a FedEx airbill number in the form at the `http://www.fedex.com` server site. A report on the shipment's progress will be displayed. It's a good application that shows how useful the World Wide Web can be for day-to-day activities. If you like the FedEx site, you might also want to try the sites for UPS (at `http://www.ups.com`) and the United States Postal Service (at `http://www.usps.gov`).

The University of Buffalo CEDAR Project address server, at `http://www.cedar.buffalo.edu/adserv.html`, features a forms-based application that looks up ZIP codes based on address information you enter into a text panel. It can also generate a downloadable PostScript label of your result (complete with a bar code) that you can print directly to an envelope, which will expedite your mail.

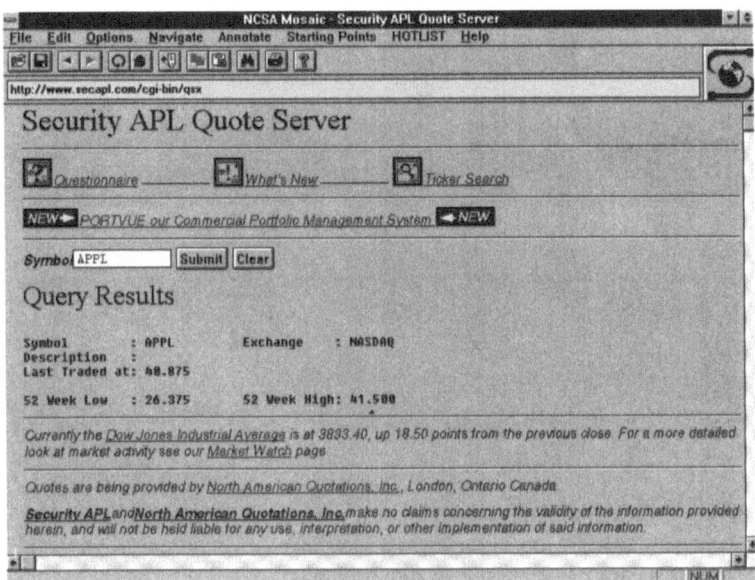

FIGURE 10.10.
The Security APL
stock quotes server.

Maps, Weather, and Geography

Weather information on the World Wide Web is divided between University sites and government installations. The Weather Server (http://thunder.atms.purdue.edu/) at Purdue University is a great place to find up-to-date satellite photos and detailed weather maps (Figure 10.11).

You can also try http://www.atmos.uiuc.edu/ for The Daily Planet from the University of Illinois at Champaign-Urbana, another good site for weather-related information.

The WebWeather site at Princeton University (http://cougarxp.princeton.edu:2112/bpd/webweather.html) puts a scrollable menu interface onto a gopher server; this makes for a very easy to use Mosaic Web browser interface. Just select an area from the menu, and a city from the next menu, and a gopher server will be contacted, and a on-time report will be transferred to your Web browser (Figure 10.12).

Government sites usually specialize in more in-depth meteorological data. A coordinating project is the National Science Foundation's UNIDATA service (http://atm.geo.nsf.gov/), a way to tie comprehensive weather data together in real time between

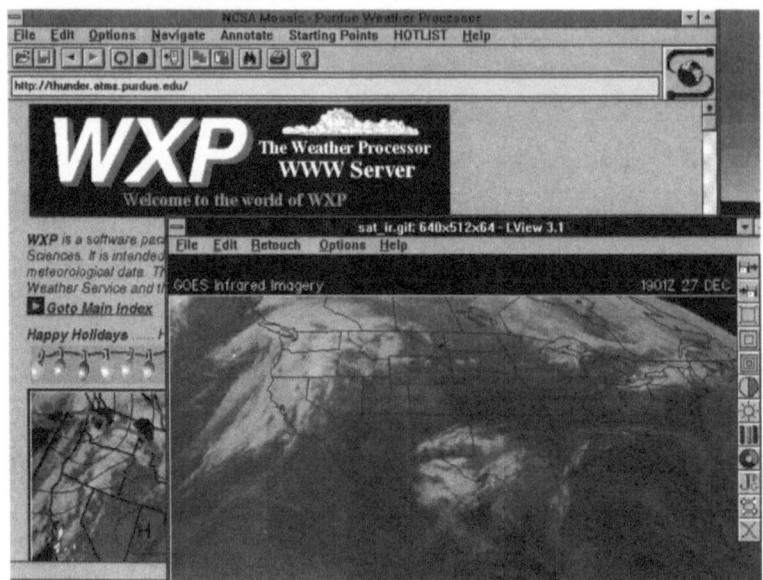

FIGURE 10.11.
The Purdue Weather
Server.

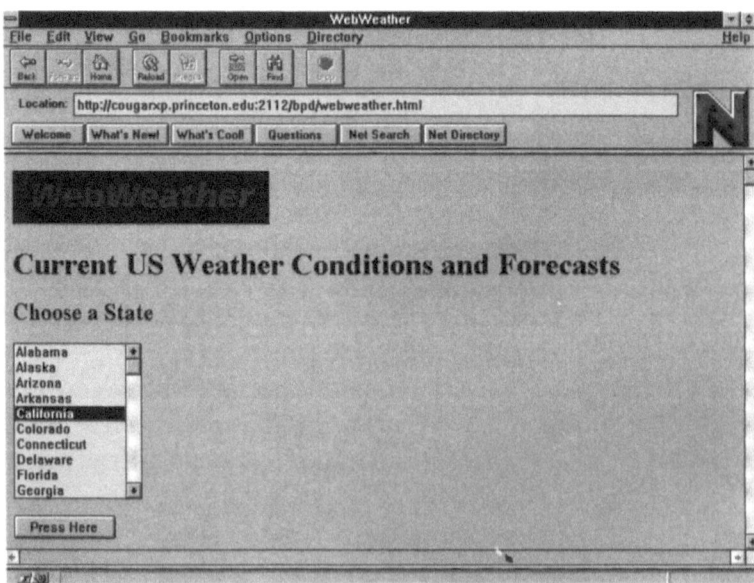

FIGURE 10.12.
WebWeather server
at Princeton
University.

sites. You can also access comprehensive weather satellite photo archives at the National Climactic Data Center (`http://www.ncdc .noaa.gov/ncdc.html`), as well as weather-related government databases.

The East and West Coast sites of the United States Geological Survey provide good environmental information. They're at `http://bramble.er.usgs.gov/` (Woods Hole USGS, Atlantic Marine Survey) and `http://walrus.wr.usgs.gov/` (Pacific Marine Geology Survey), respectively.

FIGURE 10.13.
Pacific Marine
Geology Survey.

The interactive map at the Xerox Parc (Palo Alto Research Center) is an interesting example of how the World Wide Web can display cartographic information. The map displayed can zoom in on a region by clicking on a specific area, and you can also search for a region by keywords. The Xerox map is located at `http://pubweb.parc.xerox.com/map`.

Another interesting usage of maps is located at the Southern California Department of Transportation site at `http://www. scubed.com:8001/caltrans/transnet.html`. This site shows detailed maps of traffic patterns along major roadways in Southern California and updates its maps from roadbed sensors as the day progresses, showing particular average speeds along certain stretches of road. This makes it very easy to visualize traffic jams and plan for areas to avoid. This site also provides links to map servers for other locations around the world.

Shopping

Shopping on the Internet is just getting started. With on-line credit card transaction security a real concern, you'll have to be careful about using some of the first shopping networks directly. They do offer on-line catalog interfaces that work well, however, and many of the services do business over the telephone. A representative example is the Internet Shopping Network, at `http://www.internet.net/`.

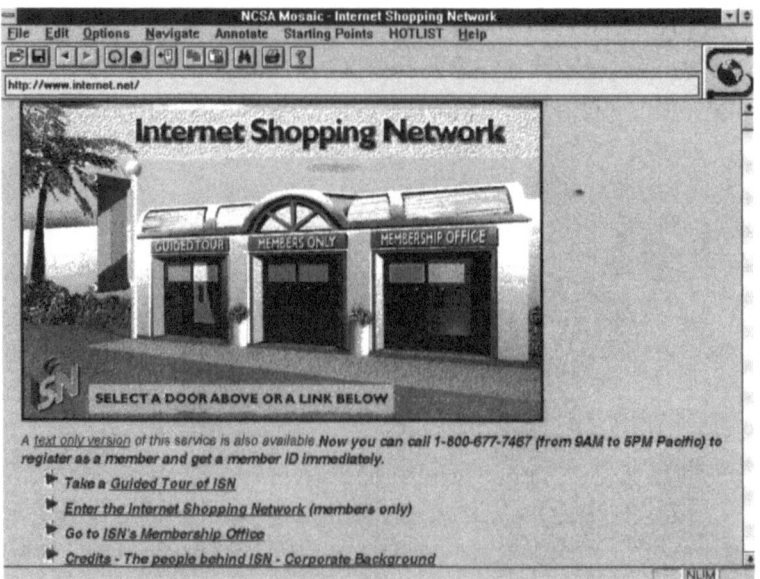

FIGURE 10.14.
The Internet
Shopping Network.

You'll also find shopping services located at major on-line sites like GNN (`http://gnn.com/gnn/GNNhome.html`), and on company home pages listed through Internet business directories like those listed previously.

11
Closing Considerations

NCSA Mosaic continues to develop and advance the public-domain version of Mosaic. The 2.0 Alpha 7 and above releases incorporate the Win32 subsystem with Microsoft's OLE (Object Linking and Embedding). The Windows Mosaic team at NCSA is working on the forefront of Windows technology, and versions for Windows 95 are already in development. It's worth noting that the latest 32-bit versions of NCSA Mosaic are ideally suited to run under Windows 95 with no modifications.

A new feature in this release is the addition of table support; this is a feature not yet found in Netscape or other Web browsers. It's a clear sign that the existence of more than one company working on Web browsers will encourage the development of competitive features, and that Mosaic will steadily improve as a result (Figure 11.1).

Netscape Communications was founded with five of the original developers of NCSA Mosaic as a commercial company, and has made a good effort to release a fast, efficient Web browser, with support for special formatting and better inline image handling. At the end of 1993, they released a 1.0 commercial version of the Netscape Web browser. This is freely available to selected groups and can be purchased directly from Netscape. Beta versions are still available on the Net, but buying the 1.0 release

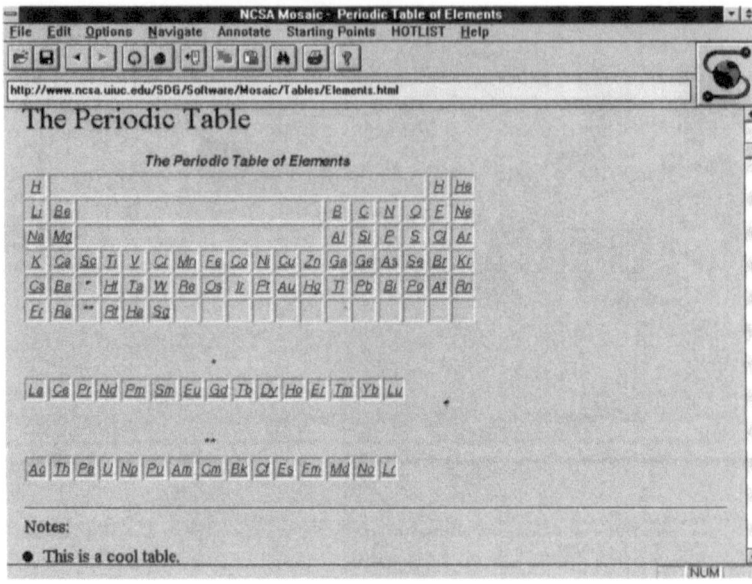

FIGURE 11.1.
The Periodic Table of Elements demo under Mosaic 2.0 Alpha 8.

entitles you to technical support from Netscape.

SPRY and Spyglass have decided to join forces and integrate AIR Mosaic with Enhanced NCSA Mosaic. This is seen as a response to the popularity of the Netscape Web browser. The combined product will have features from both Web browsers in it, and we can expect an interesting synergy to develop.

Alternate Web browsers like WebSurfer and WinWeb continue to progress, and the trend for good public-domain browsers is clearly here to stay. Microsoft has also entered the Windows Web browser fray with the beta release of Internet Assistant, an add-on to Word for Windows 6.0a.

Internet Assistant (IA) is a combination word processor/Web browser rolled into one. Upgrading your Word 6.0a installation to IA now gives you the capacity to switch between your normal document-editing view and the World Wide Web. You'll still need an Internet connection, like the one you use with Mosaic and Netscape, to access home pages with IA. It also features integrated HTML editing features, including the ability to save standard Word documents in HTML format.

As a Web browser, IA still has some ways to go before it can compete with Netscape and Mosaic. The beta features no spinning icon for tracking how fast a document is loading (although

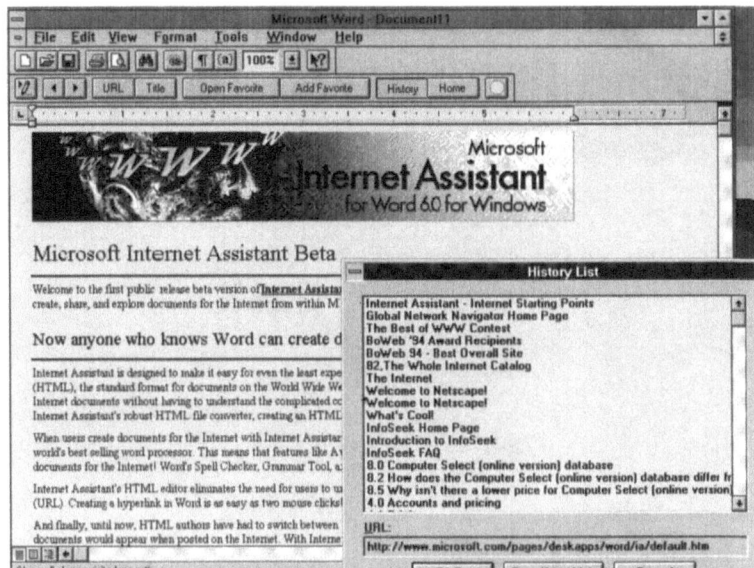

FIGURE 11.2.
Microsoft's Internet Assistant Web browser for Word for Windows 6.0a.

it does have a floating progress bar), and it doesn't handle accessing FTP sites as well as other Web browsers. But it does integrate two Windows applications into one, and it can presumably save system resources. You can now switch from editing a document in Word to a World Wide Web home page just by clicking a button, without having to launch another program.

To find out more about IA (and how to download it), use Mosaic, Netscape, or any Windows Web browser to go to `http://www.microsoft.com/pages/deskapps/word/ia/default.html`. This site features information on how to upgrade from Word 6.0 to Word 6.0a, system requirements for IA, and instructions on how to install IA for use as a part of Word for Windows.

The competition between Internet access providers should keep prices down to a reasonable level, and the availability of high-speed network equipment (like ISDN adapters) at commodity prices means that setting up a good fast connection to the Internet is well within reach for general users. The adoption of Windows standards, and the integration of Internet software into the future Win32 Windows framework, makes using a Web browser like Mosaic very easy for today's Windows users.

For a view of the path that the Internet is taking, including traffic and growth patterns, go to ftp://nic.merit.edu/nsfnet/statistics/, and you'll find a wealth of statistical data.

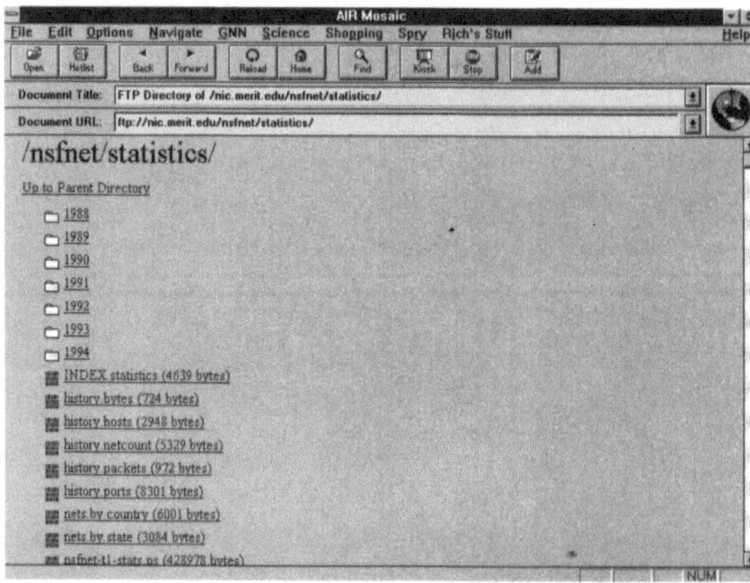

FIGURE 11.3.
The NSFNet Internet statistics FTP archive.

A World Wide Web project is developing interesting multimedia add-ons to Mosaic. These require a runtime version of Asymetrix's ToolBook program to use as a viewer. This viewer can manipulate the Mosaic program directly (by sizing it dynamically). You can also interact with items in the viewer windows (like slider bars) to change graph readouts, and manipulate items in picture views—for example, you can turn a lamp on in a picture of a dark room by clicking on it.

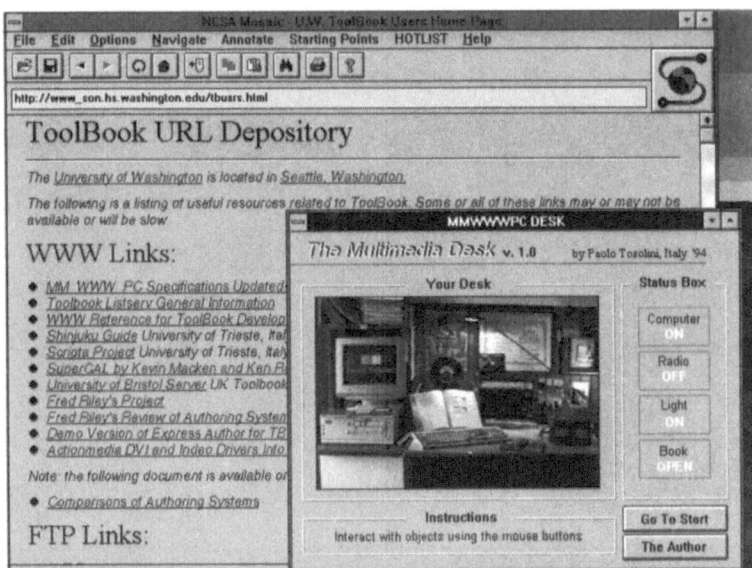

FIGURE 11.4.
Mosaic 2.0 Alpha 8 at the University of Washington ToolBook Web archive, with an interactive ToolBook program running.

A directory of resources for the ToolBook World Wide Web project, where you can find links to major Web sites, multimedia demos of various applications, and runtime software, is located at the University of Washington (http://www_son.hs.washington.edu/tbusrs.html).